Springer Theses

Recognizing Outstanding Ph.D. Research

Aims and Scope

The series "Springer Theses" brings together a selection of the very best Ph.D. theses from around the world and across the physical sciences. Nominated and endorsed by two recognized specialists, each published volume has been selected for its scientific excellence and the high impact of its contents for the pertinent field of research. For greater accessibility to non-specialists, the published versions include an extended introduction, as well as a foreword by the student's supervisor explaining the special relevance of the work for the field. As a whole, the series will provide a valuable resource both for newcomers to the research fields described, and for other scientists seeking detailed background information on special questions. Finally, it provides an accredited documentation of the valuable contributions made by today's younger generation of scientists.

Theses may be nominated for publication in this series by heads of department at internationally leading universities or institutes and should fulfill all of the following criteria

- They must be written in good English.
- The topic should fall within the confines of Chemistry, Physics, Earth Sciences, Engineering and related interdisciplinary fields such as Materials, Nanoscience, Chemical Engineering, Complex Systems and Biophysics.
- The work reported in the thesis must represent a significant scientific advance.
- If the thesis includes previously published material, permission to reproduce this must be gained from the respective copyright holder (a maximum 30% of the thesis should be a verbatim reproduction from the author's previous publications).
- They must have been examined and passed during the 12 months prior to nomination.
- Each thesis should include a foreword by the supervisor outlining the significance of its content.
- The theses should have a clearly defined structure including an introduction accessible to new PhD students and scientists not expert in the relevant field.

Indexed by zbMATH.

Xiujie Deng

Theoretical and Experimental Studies on Steady-State Microbunching

Doctoral Thesis accepted by
Tsinghua University, Beijing, China

 Springer

Author
Dr. Xiujie Deng
Department of Engineering Physics
Tsinghua University
Beijing, China

Supervisor
Prof. Wenhui Huang
Department of Engineering Physics
Tsinghua University
Beijing, China

ISSN 2190-5053   ISSN 2190-5061 (electronic)
Springer Theses
ISBN 978-981-99-5802-3   ISBN 978-981-99-5800-9 (eBook)
https://doi.org/10.1007/978-981-99-5800-9

This Springer imprint is published by the registered company Springer Nature Singapore Pte Ltd.
The registered company address is: 152 Beach Road, #21-01/04 Gateway East, Singapore 189721, Singapore

Paper in this product is recyclable.

Supervisor's Foreword

A promising accelerator light source mechanism called steady-state microbunching (SSMB) has been actively studied in recent years. The idea of SSMB is to scale the longitudinal focusing of the electron beam in a storage ring from the conventional radiofrequency range to optical laser wavelengths. The combination of microbunching-enabled coherent radiation and the high repetition rate of electron beam circulating in a storage ring makes SSMB a high-average-power high-flux narrowband photon source, with wavelength extendable to soft X-ray. Such a high-flux narrowband light source allows sub-meV energy resolution in angle-resolved photoemission spectroscopy, which could provide new opportunities for fundamental physics research, for example, to probe the energy gap distribution and electronic states of superconducting materials. An SSMB-based kW-level extreme ultraviolet (EUV) source is also appealing for the semiconductor industry to be used in EUV lithography for high-volume chip manufacturing.

From the accelerator physics perspective, the six orders of magnitude shortening of bunch length compared to that in a conventional storage ring provide challenges as well as tremendous opportunities for the accelerator physics research of SSMB. The Ph.D. thesis of Dr. Xiujie Deng is devoted to the theoretical and experimental studies of SSMB, with important results achieved. The contribution of his thesis can be summarized into three categories: first, answer the question of how to realize SSMB; second, reveal what radiation characteristics can be obtained from the formed SSMB; and third, experimentally demonstrate the working mechanism of SSMB in a real machine for the first time. All these achievements, in particular, the first proof-of-principle experiment, are of crucial and fundamental importance for the development of SSMB. In addition, I believe the efforts of Dr. Deng on precision microbunching dynamics will be of growing value to the general accelerator community, as the requirement for beam manipulation becomes more and more demanding. I wholeheartedly recommend this thesis to all accelerator scientists, especially those who work on advanced light sources, and also to the potential users of SSMB.

Beijing, China
May 2023

Wenhui Huang

Parts of this thesis have been published in the following documents

Journal Publications

Deng, X., et al. Experimental demonstration of the mechanism of steady-state microbunching. Nature 590.7847 (2021): 576–579.

Deng, X. J., et al. Average and statistical property of coherent radiation from steady-state microbunching. J. Synchrotron Rad. (2023). 30, 35–50.

Deng, X. J., et al. Breakdown of classical bunch length and energy spread formula in a quasi-isochronous electron storage ring. Phys. Rev. Accel. Beams 26.5 (2023): 054001.

Deng, X. J., et al. Courant-Snyder formalism of longitudinal dynamics. Phys. Rev. Accel. Beams 24.9 (2021): 094001.

Deng, X. J., et al. Single-particle dynamics of microbunching. Phys. Rev. Accel. Beams 23.4 (2020): 044002.

Deng, X. J., et al. Widening and distortion of the particle energy distribution by chromaticity in quasi-isochronous rings. Phys. Rev. Accel. Beams 23.4 (2020): 044001.

Deng, X. J., et al. Harmonic generation and bunch compression based on transverse-longitudinal coupling. Nucl. Instrum. Methods Phys. Res. A 1019 (2021): 165859.

Tang, C. & Deng, X. Steady-State Micro-Bunching Accelerator Light Source. Acta Phys. Sin., 2022, 71(15): 152901.

Zhang, Y., et al. Ultralow longitudinal emittance storage rings. Phys. Rev. Accel. Beams 24.9 (2021): 090701.

Acknowledgements

This dissertation is dedicated to my wife Congcong Xu, my mother Jinping Yang, and my father Hong Deng. The completion of it is impossible without the support of many colleagues. Please excuse me for listing only part of them here due to limited space.

First, I want to give my great gratitude to Prof. Alex Chao who has reawaken my passion for physics and set an example as a physicist. The development of SSMB owes a lot to the insights and promotion of Prof. Chao. Much of my accelerator physics knowledge attributes to my study with Prof. Chao. I always treasure the important encouragements from Prof. Chao when I needed them most to build my confidence as a researcher.

Much appreciation goes to my supervisor Prof. Wenhui Huang, who has provided me with the precious opportunity to join the SSMB research and continuous guidance. Professor Huang has given me flexibility and support beyond what I could have imagined. I always feel touched by the fact that Prof. Huang is happy when I made achievements in science or life.

I also thank Prof. Chuanxiang Tang very much for providing the opportunity, continuous guidance, and support in the past years. Without the decision of Prof. Tang and Prof. Chao to initiate the SSMB research at Tsinghua University, this thesis will not be here. Much achievement of the Tsinghua SSMB task force is only possible with the lead of Prof. Tang.

Special thanks goes to Dr. Jörg Feikes, for the exciting work together on SSMB experiments at the MLS, the joyful discussions on physics and the other, the appreciation of me, and various help during and after my Ph.D. training. The experience at the MLS gives me a chance to appreciate the complexity, richness, and robustness of a real machine.

The efforts of the SSMB proof-of-principle experiment group and the numerous support from Tsinghua, HZB, and PTB colleagues are also much appreciated. This research is supported by the National Natural Science Foundation of China (NSFC Grant No. 12035010) and Shuimu Tsinghua Scholar Program of Tsinghua University.

Contents

1 **Introduction** 1
   References 4

2 **SSMB Longitudinal Dynamics** 7
   2.1 Linear Longitudinal Dynamics 8
        2.1.1 Longitudinal Courant-Snyder Formalism 8
        2.1.2 Classical $\sigma_z \propto \sqrt{|\eta|}$ Scaling 12
        2.1.3 Beyond the Classical $\sigma_z \propto \sqrt{|\eta|}$ Scaling 14
        2.1.4 Campbell's Theorem 23
        2.1.5 Minimizing Longitudinal Emittance 25
        2.1.6 Longitudinal Strong Focusing 34
        2.1.7 Thick-Lens Maps of a Laser Modulator 37
   2.2 Nonlinear Longitudinal Dynamics 40
        2.2.1 For High-Harmonic Bunching 40
        2.2.2 For Longitudinal Dynamic Aperture 43
   References 47

3 **SSMB Transverse-Longitudinal Coupling Dynamics** 51
   3.1 Linear Transverse-Longitudinal Coupling Dynamics 52
        3.1.1 Passive Bunch Lengthening 52
        3.1.2 Coupling for Harmonic Generation and Bunch
               Compression 56
        3.1.3 Normal RF or TEM00 Mode Laser for Coupling 61
        3.1.4 Transverse Deflecting RF or TEM01 Mode Laser
               for Coupling 68
        3.1.5 Emittance Exchange 72
   3.2 Nonlinear Transverse-Longitudinal Coupling Dynamics 76
        3.2.1 Average Path Length Dependence on Betatron
               Amplitudes 76
        3.2.2 Energy Widening and Distortion 77
        3.2.3 Experimental Verification 80
   References 85

4   SSMB Radiation 89
  4.1   Formulation of Radiation from a 3D Rigid Beam 90
  4.2   Form Factors 92
    4.2.1   Longitudinal Form Factor 93
    4.2.2   Transverse From Factor 95
  4.3   Radiation Power and Spectral Flux 101
  4.4   Impact of Electron Beam Divergence and Energy Spread 104
  4.5   Statistical Property of Radiation 105
    4.5.1   Pointlike Nature of Electron 106
    4.5.2   Quantum Nature of Radiation 111
  4.6   Example Calculations for Envisioned EUV SSMB 112
    4.6.1   Average Property 113
    4.6.2   Statistical Property 115
    4.6.3   Discussions 116
  References 117

5   SSMB Proof-of-Principle Experiments 119
  5.1   Strategy of the PoP Experiments 119
    5.1.1   Three Stages of PoP Experiments 119
    5.1.2   Metrology Light Source Storage Ring 121
  5.2   PoP I: Turn-by-Turn Laser-Electron Phase Correlation 122
    5.2.1   Experimental Setup 122
    5.2.2   Physical Analysis of Microbunching Formation 123
    5.2.3   Microbunching Radiation Calculation 130
    5.2.4   Signal Detection and Evaluation 134
    5.2.5   Experimental Results 137
    5.2.6   Summary 141
  5.3   PoP II: Quasi-steady-State Microbunching 142
    5.3.1   Phase-Mixing in Buckets 142
    5.3.2   Experimental Parameters Choice 147
  References 148

6   Summary 151
  6.1   Summary of the Dissertation 151
  6.2   Useful Formulas and Example Parameters for SSMB Storage
    Rings 153
    6.2.1   Longitudinal Weak Focusing SSMB 153
    6.2.2   Transverse-Longitudinal Coupling SSMB 156
  References 160

Chapter 1
Introduction

Particle accelerators as photon sources are advanced tools in investigating the structure and dynamical properties of matter, and have enabled advances in science and technology for more than half a century. The present workhorses of these sources are storage ring-based synchrotron radiation facilities [1–3] and linear accelerator-based free-electron lasers (FELs) [4–7]. These two kinds of sources deliver light with high repetition rate and high peak brilliance and power, respectively. Some applications, however, do need high average power and high photon flux. Kilowatt extreme ultraviolet (EUV) light sources, for example, are urgently needed by the semiconductor industry for EUV lithography [8]. Another example is that to realize high energy resolution in synchrotron-based angle-resolved photoemission spectroscopy (ARPES), which is highly desired by fundamental condensed matter physics research, we need the initial radiation photon flux before monochromator is high enough. To obtain high average power and high photon flux, a high peak power or a high repetition rate alone is not sufficient. We need both of them simultaneously.

The key of the high peak power of FELs lies in microbunching, which means the electrons are bunched or sub-bunched to a longitudinal dimension smaller than the radiation wavelength so that the electrons radiate in phase and thus cohere [9–11]. The power of coherent radiation is proportional to the number of the radiating electrons squared, therefore can be orders of magnitude stronger than the equivalent incoherent radiation in which the power dependence on the electron number is linear. The Self-Amplified Spontaneous Emission (SASE) scheme [6, 7] of microbunching making the high-gain FELs so powerful, however, is actually a collective beam instability which degrades the electron beam parameters and the microbunching can only be exploited once. The repetition rate of the radiation is thus limited by the repetition rate of the driving source, i.e., the linear accelerator. There are now active efforts devoted to improve the repetition rate of FEL radiation, for example by implementing the superconducting technology. However, the realization of a high-average-power, continuous-wave (CW), narrowband, short-wavelength light source remains a challenge.

© The Author(s) 2024
X. Deng, *Theoretical and Experimental Studies on Steady-State Microbunching*,
Springer Theses, https://doi.org/10.1007/978-981-99-5800-9_1

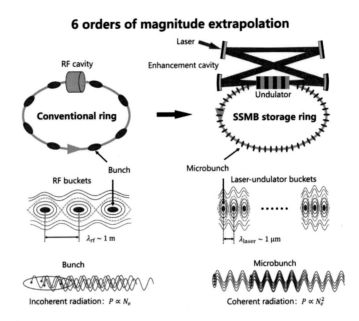

Fig. 1.1 A schematic layout of a conventional storage ring (left) and an SSMB storage ring (right)

A mechanism called steady-state microbunching (SSMB) has been proposed [12, 13] to resolve this issue. The idea of SSMB is that by a phase-space manipulation of an electron beam, microbunching forms and stays in a steady state each time going through a radiator in a storage ring. The steady state here means a balance of excitation and damping, a true equilibrium in the context of electron storage ring beam dynamics. The schematic layout of an SSMB storage ring and its operating principle in comparison to a conventional storage ring is shown in Fig. 1.1. SSMB replaces the conventional bunching system in a storage ring, namely the radiofrequency (RF) cavity, with a laser modulation system. As the wavelength of laser ($\sim \mu$m) is typically six orders of magnitude smaller than that of an RF wave ($\sim$ m), a much shorter bunch, i.e., microbunch, can thus be anticipated by invoking this replacement together with a dedicated storage ring magnetic lattice.

The microbunching in SSMB is from the active longitudinal focusing provided by the laser modulator, just similar to the conventional RF bunching through phase stability principle [14, 15]. The radiation in SSMB, unlike that in an FEL, is a passive process and the radiator can be rather short, for example it can be a simple dipole magnet or a short undulator. The SSMB modulator is also much shorter than the radiator undulator in a high-gain FEL. Therefore, there is no FEL mechanism invoked in the bunching or radiation process in SSMB. If there is some unavoidable FEL effects, it needs to be controlled within a safe region to not destroy the steady state micobunches.

To provide adequate and stable longitudinal focusing such that microbunches can be formed and sustained, SSMB requires a powerful phase-locked laser to interact with electrons on a turn-by-turn basis. The realization of such a laser system usually demands an optical enhancement cavity. A laser cannot effectively interact with the co-propagating electrons if the electrons go through a straight line, as the electric field of a laser is perpendicular to the laser propagation direction. A modulator which bends the electron trajectory transversely is thus needed. The modulator is usually an undulator, which is a periodic structure of dipole magnets with oscillating polarity. Note that to avoid the head-on collisions, i.e., the Compton back-scattering, between the reflected laser and the electrons, a four-mirror optical cavity, instead of a two-mirror one, is chosen for the illustration in Fig. 1.1.

Note that we have not presented explicitly the energy replenish system for SSMB in the illustration. The modulation laser in principle can be used to compensate the radiation energy loss of the electrons, just like the traditional RF, but this may not be a cost-effective choice. Besides, the electron beam current and output radiation power will also be limited by the incident laser power. Instead, one may just use a traditional RF cavity for the energy compensation. If a larger filling factor of the electron beam is desired, the energy supply system could also be one or several induction acceleration cavities. In the present envisioned high-average-power SSMB photon source, induction linac is tentatively used as the energy compensation system and the filling factor of the electron beam in the storage ring can be rather large, for example larger than 50%.

Once realized, SSMB can combine the strong coherent radiation from microbunching and the high repetition rate of beam circulating in a storage ring to provide high-average-power, high-repetition (MHz to CW) narrowband radiation, with the wavelength ranging from THz to soft X-ray. Such a novel photon source could provide unprecedented opportunities for accelerator photon science and technological applications. For example, SSMB is promising for generating kW-level EUV radiation for EUV lithography [16]. Energy-tunable high-flux narrowband EUV photons are also highly desirable in condensed matter physics study, such as used in high-resolution ARPES to probe the energy gap distribution and electronic states of superconducting materials. Ultrahigh-power deep ultraviolet and infrared sources are potential research tools in atomic and molecular physics. Moreover, new nonlinear phenomena and dynamical properties of materials can be driven and studied by high-peak and average-power THz sources. Besides high power, SSMB can also produce ultrashort (sub-femtosecond to attosecond) photon pulse trains with definite phase relations, which could be useful in attosecond physics investigations.

This dissertation is devoted to the theoretical and experimental studies of SSMB, with important results achieved. The contribution of this dissertation can be summarized as: first, answer the question of how to realize SSMB; second, reveal what radiation characteristics can we obtain from the formed SSMB; and third, experimentally demonstrate the working mechanism of SSMB in a real machine for the first time. More specifically, in Chaps. 2 and 3, we have conducted in-depth theoretical and experimental studies on single-particle effects vital for the formation and transportation of microbunching in a storage ring. Chapter 2 is on longitudinal dynamics, while Chap. 3 is devoted to transverse-longitudinal coupling dynamics. Chapter 4 is the theoretical and numerical investigation on the average and statistical characteristics of the radiation generated from the formed microbunching. In Chap. 5, we report our work on the first successful demonstration of the mechanism of SSMB, performed at the Metrology Light Source in Berlin. Finally, in Chap. 6 we present a short summary of the dissertation, together with some useful formulas and example parameters of SSMB storage rings aimed for kW-level infrared, EUV and soft X-ray radiation, respectively. Summarizing, the highlights of this dissertation are:

- Presents the first proof-of-principle experiment of a promising accelerator light source mechanism.
- Covers precision longitudinal and transverse-longitudinal coupling dynamics in a storage ring.
- Provides useful formulas and example parameters for high-power infrared, EUV and soft X-ray light source design.

The work presented in this dissertation is of fundamental importance for the development of an SSMB-based high-power photon source.

References

1. Elder FR, Gurewitsch AM, Langmuir RV, Pollock HC (1947) Radiation from electrons in a synchrotron. Phys Rev 71:829–830
2. Tzu HY (1948) On the radiation emitted by a fast charged particle in the magnetic field. Proc R Soc Lond A 192(1029):231–246
3. Schwinger J (1949) On the classical radiation of accelerated electrons. Phys Rev 75(12):1912
4. Madey JM (1971) Stimulated emission of bremsstrahlung in a periodic magnetic field. J Appl Phys 42(5):1906–1913
5. Kroll NM, McMullin WA (1978) Stimulated emission from relativistic electrons passing through a spatially periodic transverse magnetic field. Phys Rev A 17:300–308
6. Kondratenko A, Saldin E (1980) Generating of coherent radiation by a relativistic electron beam in an ondulator. Part Accel 10:207–216
7. Bonifacio R, Pellegrini C, Narducci L (1984) Collective instabilities and high-gain regime free electron laser. In: AIP conference proceedings, vol 118, pp 236–259. American Institute of Physics
8. Bakshi V (2018) Euv lithography. In: Society of photo-optical instrumentation engineers (SPIE)
9. Schwinger J (1996) On radiation by electrons in a betatron. In: A quantum legacy: seminal papers of Julian Schwinger. World Scientific, pp 307–331
10. Nodvick JS, Saxon DS (1954) Suppression of coherent radiation by electrons in a synchrotron. Phys Rev 96(1):180
11. Gover A, Ianconescu R, Friedman A, Emma C, Sudar N, Musumeci P, Pellegrini C (2019) Superradiant and stimulated-superradiant emission of bunched electron beams. Rev Mod Phys 91:035003
12. Ratner DF, Chao AW (2010) Steady-state microbunching in a storage ring for generating coherent radiation. Phys Rev Lett 105:154801
13. Chao A, Granados E, Huang X, Ratner D, Luo H-W (2016) High power radiation sources using the steady-state microbunching mechanism. In: Proceedings of the 7th international particle accelerator conference (IPAC'16), Busan, Korea, 2016. JACoW, Geneva, pp 1048–1053

14. Veksler VI (1944) New method for the acceleration of relativistic particles. Doklady Akademii Nauk USSR 43:346–348
15. McMillan EM (1945) The synchrotron-a proposed high energy particle accelerator. Phys Rev 68(5–6):143
16. Tang C, Deng X (2022) Steady-state micro-bunching accelerator light source. Acta Phys Sin 71:152901

Chapter 2
SSMB Longitudinal Dynamics

In this chapter, we study the single-particle longitudinal dynamics of SSMB. The motivation is to answer the question: how to realize the short bunch length and small longitudinal emittance in an electron storage ring, as required by SSMB? Note that the curvilinear (Frenet-Serret) coordinate system and the state vector $\mathbf{X} = (x, x', y,' y', z, \delta)^T$, with ^{T} representing the transpose, are used throughout this dissertation. For the longitudinal dynamics without coupling from the transverse dimension, what we can play are the momentum compaction and RF systems, for SSMB the laser modulators. The momentum compaction is a measure of particle energy dependence of the recirculation path length

$$\alpha = \frac{\Delta C/C_0}{\Delta E/E_0} = \frac{1}{C_0} \oint \frac{D_x(s)}{\rho(s)} ds, \tag{2.1}$$

where C_0 is the ring circumference, E_0 is the particle energy, D_x is the horizontal dispersion which is a measure of the energy dependence of particle horizontal position, ρ is the bending radius. Considering the energy-dependent velocity, the particle energy dependence of the revolution time can be quantified by a parameter named phase slippage factor

$$\eta = \frac{\Delta T/T_0}{\Delta E/E_0} = \alpha - \frac{1}{\gamma^2}, \tag{2.2}$$

with γ the Lorentz factor. For linear dynamics, the phase slippage to longitudinal dimension is like the drift space to transverse dimension, while the RF kick in linear approximation to longitudinal dimension is like the quadrupole to transverse dimension. The difference is that the sign of phase slippage can either be positive or negative, while the drift space can only have a positive physical length. To account for the impact of local or partial phase slippage on the evolution of longitudinal optics around the ring, Courant-Snyder analysis can be invoked for linear dynamics study beyond adiabatic approximation. Such new derivations are necessary to accurately describe the dynamics of the SSMB mechanism. Usually there is only one RF cavity

© The Author(s) 2024
X. Deng, *Theoretical and Experimental Studies on Steady-State Microbunching*, Springer Theses, https://doi.org/10.1007/978-981-99-5800-9_2

in a storage ring, the longitudinal optics can be manipulated with more freedom with multiple RFs. For example, the strong focusing principle can be implemented in the longitudinal dimension to realize ultrashort bunch length, not unlike its transverse counterpart. For nonlinear dynamics, both the nonlinearity of the phase slippage and the sinusoidal modulation waveform can lead to subtle and rich beam dynamics. In the following we will investigate along this brief review. Parts of the work presented in this chapter have been published in Refs. [1–4].

2.1 Linear Longitudinal Dynamics

2.1.1 Longitudinal Courant-Snyder Formalism

SSMB means an ultrashort electron bunch in an equilibrium state. One successful method of realizing short bunches in an electron storage ring is the implementation of a quasi-isochronous lattice, which means particles with different energies complete one revolution using almost the same time. The reason behind is the well-known $\sigma_z \propto \sqrt{|\eta|}$ scaling law of the "zero-current" bunch length given by Sands [5], in which $\eta = \alpha - \frac{1}{\gamma^2}$ is the global phase slippage factor of the ring as introduced just now. However, from single-particle dynamics perspective, there is a fundamental effect limiting the lowest bunch length realizable in an electron storage ring originating from the stochasticity of photon emission time or location. This stochasticity results in a diffusion of the electron longitudinal coordinate z even if the global phase slippage of the ring is zero as we cannot make all the local or partial phase slippages zero simultaneously. The partial phase slippage factor from s_1 to s_2 is defined as

$$\tilde{\eta}(s_2, s_1) = \frac{1}{C_0} \int_{s_1}^{s_2} \left(\frac{D_x(s)}{\rho(s)} - \frac{1}{\gamma^2} \right) ds. \tag{2.3}$$

The physical picture of the partial phase slippage and quantum excitation in both the particle energy and longitudinal coordinate, therefore the longitudinal emittance, is shown in Fig. 2.1.

Due to this quantum diffusion, there exists a lower bunch length limit and the energy spread diverges when the bunch length is pushed close to the limit. This effect is of vital importance for SSMB and other ideas invoking ultrashort electron bunches or ultrasmall longitudinal emittance in storage rings. It is first theoretically investigated by Shoji et al. [6, 7], and recently more accurately analyzed by us using the longitudinal Courant-Snyder formalism [2, 3, 8]. The key to understanding the effect is to change from the global viewpoint to a local one, i.e., the quantum excitation at different places around the ring actually contribute to the longitudinal emittance with different strengths, just like its transverse counterpart.

For an accurate analysis of this effect, here we invoke Chao's solution by linear matrices (SLIM) formalism [9]. SLIM is an early effort to generalize the classical

Fig. 2.1 Physical picture of the partial phase slippage factors and quantum excitation. Particles undergo diffusion in both the particle energy and the longitudinal coordinate in each turn, giving rise to longitudinal emittance growth

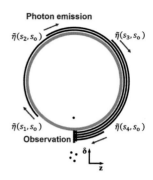

Courant-Snyder theory [10] from 1D (2D phase space) to higher dimensions. It invokes 6×6 transport matrices and applies to 3D (6D phase space) general coupled lattice without the assumption of a small synchrotron tune. Concerning the evaluation of equilibrium beam parameters in an electron storage ring, SLIM can be viewed as a method of solving linear Fokker-Planck equation [11, 12] without adopting the adiabatic approximation. Therefore, it can account for the variation of one-turn map around the ring. In other words, the contribution of diffusion and damping, for example the quantum excitation and radiation damping, to the eigen emittances depends on the local one-turn map.

The three eigen emittances ϵ_k of a particle beam, with $k = I, II, III$, are defined as the positive eigenvalues of $i\,\mathbf{\Sigma S}$, where i is the imaginary unit, $\mathbf{\Sigma} = \langle \mathbf{XX}^T \rangle$ are the second moments of the beam and

$$\mathbf{S} = \begin{pmatrix} 0 & 1 & 0 & 0 & 0 & 0 \\ -1 & 0 & 0 & 0 & 0 & 0 \\ 0 & 0 & 0 & 1 & 0 & 0 \\ 0 & 0 & -1 & 0 & 0 & 0 \\ 0 & 0 & 0 & 0 & 0 & 1 \\ 0 & 0 & 0 & 0 & -1 & 0 \end{pmatrix}. \tag{2.4}$$

The eigen emittances are invariants with respect to a linear symplectic transportation $\mathbf{T}$, as

$$i\,\mathbf{\Sigma}_{\text{new}}\mathbf{S} = i\mathbf{T}\mathbf{\Sigma}_{\text{old}}\mathbf{T}^T\mathbf{S} = \mathbf{T}(i\,\mathbf{\Sigma}_{\text{old}}\mathbf{S})\mathbf{T}^{-1}, \tag{2.5}$$

in which the last step has invoked the symplecticity of $\mathbf{T}$, i.e., $\mathbf{T}^T\mathbf{ST} = \mathbf{S}$. Therefore, $i\,\mathbf{\Sigma}_{\text{new}}\mathbf{S}$ is related to $i\,\mathbf{\Sigma}_{\text{old}}\mathbf{S}$ by a similarity transform, thus having the same eigenvalues.

In an electron storage ring, the equilibrium state is a balance between quantum excitation and radiation damping. According to SLIM [9], the equilibrium eigen emittances are given by

$$\epsilon_k = \frac{C_L \gamma^5}{c\alpha_k} \oint \frac{|\mathbf{E}_{k5}(s)|^2}{|\rho(s)|^3} ds, \tag{2.6}$$

and the second moments of the beam are

$$\Sigma_{ij} = 2 \sum_{k=I,II,III} \epsilon_k \mathrm{Re}[\mathbf{E}_{ki}\mathbf{E}_{kj}^*],$$

(2.7)

where α_k are the damping constants of the three eigen modes, $C_L = 55 r_e \hbar /$ $(48\sqrt{3}m_e)$, with $\hbar$ the reduced Planck's constant, r_e the electron classical radius, Re[] means taking the real part of a complex number, * means complex conjugate, and $\mathbf{E}_k$ are eigenvectors of the 6×6 symplectic one-turn map, satisfying the following normalization condition

$$\mathbf{E}_k^{\dagger}\mathbf{S}\mathbf{E}_k = \begin{cases} i, & k = I, II, III, \\ -i, & k = -I, -II, -III, \end{cases}$$

(2.8)

and $\mathbf{E}_k^{\dagger}\mathbf{S}\mathbf{E}_j = 0$ for $k \neq j$, in which † means complex conjugate transpose. $\mathbf{E}_{ki}$ is the i-th component of the eigenvector $\mathbf{E}_k$.

To simplify the discussion, here we only consider the horizontal and longitudinal dimensions and use the state vector $\mathbf{X} = (x, x', z, \delta)^T$. Under the assumptions that the ring is planar x-y uncoupled and the RFs are placed at dispersion-free locations, which is the typical setup for present synchrotron radiation sources, the betatron coordinate $\mathbf{X}_\beta = \mathbf{B}\mathbf{X}$ can be introduced to parametrize the transfer matrix in a diagonal form, with the dispersion matrix given by

$$\mathbf{B} = \begin{pmatrix} 1 & 0 & 0 & -D_x \\ 0 & 1 & 0 & -D_x' \\ D_x' & -D_x & 1 & 0 \\ 0 & 0 & 0 & 1 \end{pmatrix}.$$

(2.9)

The one-turn map $\mathbf{M}$ of $\mathbf{X}$ is related to the one-turn map $\mathbf{M}_\beta$ of $\mathbf{X}_\beta$ by

$$\mathbf{M} = \mathbf{B}^{-1}\mathbf{M}_\beta\mathbf{B},$$

(2.10)

with

$$\mathbf{M}_\beta = \begin{pmatrix} \mathbf{M}_{x\beta} & 0 \\ 0 & \mathbf{M}_{z\beta} \end{pmatrix},$$

$$\mathbf{M}_{x,z\beta} = \begin{pmatrix} \cos\Phi_{x,z} + \alpha_{x,z}\sin\Phi_{x,z} & \beta_{x,z}\sin\Phi_{x,z} \\ -\gamma_{x,z}\sin\Phi_{x,z} & \cos\Phi_{x,z} - \alpha_{x,z}\sin\Phi_{x,z} \end{pmatrix},$$

(2.11)

in which $\Phi_x = 2\pi\nu_x$ and $\Phi_z = 2\pi\nu_s$ are the betatron and synchrotron phase advance per turn. The eigenvectors of $\mathbf{M}_\beta$ can be expressed using the Courant-Snyder functions as

$$\mathbf{v}_x = \frac{1}{\sqrt{2}} \begin{pmatrix} \sqrt{\beta_x} \\ \frac{i-\alpha_x}{\sqrt{\beta_x}} \\ 0 \\ 0 \end{pmatrix} e^{i\Phi_I}, \quad \mathbf{v}_z = \frac{1}{\sqrt{2}} \begin{pmatrix} 0 \\ 0 \\ \sqrt{\beta_z} \\ \frac{i-\alpha_z}{\sqrt{\beta_z}} \end{pmatrix} e^{i\Phi_{III}}, \tag{2.12}$$

where Φ_I and Φ_{III} are phase factors which do not affect the normalization of eigenvector and the calculation of physical quantities. Therefore, the eigenvectors of $\mathbf{M}$ are

$$\mathbf{E}_I = \mathbf{B}^{-1}\mathbf{v}_x = \frac{1}{\sqrt{2}} \begin{pmatrix} \sqrt{\beta_x} \\ \frac{i-\alpha_x}{\sqrt{\beta_x}} \\ -\sqrt{\beta_x}D_x' + \frac{i-\alpha_x}{\sqrt{\beta_x}}D_x \\ 0 \end{pmatrix} e^{i\Phi_I}, \quad \mathbf{E}_{III} = \mathbf{B}^{-1}\mathbf{v}_z = \frac{1}{\sqrt{2}} \begin{pmatrix} \frac{i-\alpha_z}{\sqrt{\beta_z}}D_x \\ \frac{i-\alpha_z}{\sqrt{\beta_z}}D_x' \\ \sqrt{\beta_z} \\ \frac{i-\alpha_z}{\sqrt{\beta_z}} \end{pmatrix} e^{i\Phi_{III}}.$$
$$\tag{2.13}$$

According to SLIM, the equilibrium horizontal and longitudinal emittance are then

$$\begin{aligned} \epsilon_x &\equiv \langle J_x \rangle = \frac{55}{96\sqrt{3}} \frac{\alpha_F \lambdabar_e^2 \gamma^5}{\alpha_H} \oint \frac{\mathcal{H}_x(s)}{|\rho(s)|^3} ds, \\ \epsilon_z &\equiv \langle J_z \rangle = \frac{55}{96\sqrt{3}} \frac{\alpha_F \lambdabar_e^2 \gamma^5}{\alpha_L} \oint \frac{\beta_z(s)}{|\rho(s)|^3} ds, \end{aligned} \tag{2.14}$$

in which

$$\begin{aligned} J_x &= \frac{(x - D_x\delta)^2 + [\alpha_x(x - D_x\delta) + \beta_x(x' - D_x'\delta)]^2}{2\beta_x}, \\ J_z &= \frac{(z - D_x'x - D_xx')^2 + [\alpha_z(z - D_x'x - D_xx') + \beta_z\delta]^2}{2\beta_z}, \end{aligned} \tag{2.15}$$

are the horizontal and longitudinal action of a particle, and $\langle \rangle$ here means particle ensemble average, α_H and α_L are the horizontal and longitudinal damping constants,

$$\alpha_H = \frac{U_0}{2E_0}(1 - \mathcal{D}), \quad \alpha_L = \frac{U_0}{2E_0}(2 + \mathcal{D}), \tag{2.16}$$

where U_0 is the radiation energy loss of a particle per turn, $\mathcal{D} = \frac{\oint \frac{(1-2n)D_x}{\rho^3} ds}{\oint \frac{1}{\rho^2} ds}$, with $n = -\frac{\rho}{B}\frac{\partial B}{\partial \rho}$ the field gradient index, $\alpha_F = \frac{1}{137}$ is the fine-structure constant, $\lambdabar_e = \lambda_e/2\pi = 386$ fm is the reduced Compton wavelength of electron and $\mathcal{H}_x = \gamma_x D_x^2 + 2\alpha_x D_x D_x' + \beta_x D_x'^2 = \frac{D_x^2 + (\alpha_x D_x + \beta_x D_x')^2}{\beta_x}$ is the horizontal chromatic function. Therefore, it is the longitudinal beta function β_z at the bending magnets that matters in determining the contribution of quantum excitation to the longitudinal emittance ϵ_z. A physical picture is given in Fig. 2.2 to help better understand this argument.

Fig. 2.2 A physical picture
to explain why a larger
longitudinal beta function β_z
means a larger contribution
to longitudinal emittance ϵ_z,
with a given strength of
quantum excitation

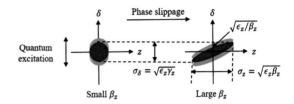

After getting the equilibrium eigen emittances, we can obtain the second moments
of the beam according to Eqs. (2.7) and (2.13), more specifically,

$$
\Sigma_\beta = \langle \mathbf{X}_\beta \mathbf{X}_\beta^T \rangle =
\begin{pmatrix}
\epsilon_x \beta_x & -\epsilon_x \alpha_x & 0 & 0 \\
-\epsilon_x \alpha_x & \epsilon_x \gamma_x & 0 & 0 \\
0 & 0 & \epsilon_z \beta_z & -\epsilon_z \alpha_z \\
0 & 0 & -\epsilon_z \alpha_z & \epsilon_z \gamma_z
\end{pmatrix},
\tag{2.17}
$$

and

$$
\Sigma = \langle \mathbf{X}\mathbf{X}^T \rangle = \mathbf{B}^{-1} \Sigma_\beta \left(\mathbf{B}^{-1} \right)^T = \begin{pmatrix} \Sigma_H & \Sigma_{HL} \\ \Sigma_{HL}^T & \Sigma_L \end{pmatrix},
\tag{2.18}
$$

where

$$
\begin{aligned}
\Sigma_H &= \begin{pmatrix} \epsilon_x \beta_x + \epsilon_z \gamma_z D_x^2 & -\epsilon_x \alpha_x + \epsilon_z \gamma_z D_x D_x' \\ -\epsilon_x \alpha_x + \epsilon_z \gamma_z D_x D_x' & \epsilon_x \gamma_x + \epsilon_z \gamma_z D_x'^2 \end{pmatrix}, \\
\Sigma_{HL} &= \begin{pmatrix} -\epsilon_x(\alpha_x D_x + \beta_x D_x') - \epsilon_z \alpha_z D_x & \epsilon_z \gamma_z D_x \\ \epsilon_x(\gamma_x D_x + \alpha_x D_x') - \epsilon_z \alpha_z D_x' & \epsilon_z \gamma_z D_x' \end{pmatrix}, \\
\Sigma_L &= \begin{pmatrix} \epsilon_x \mathcal{H}_x + \epsilon_z \beta_z & -\epsilon_z \alpha_z \\ -\epsilon_z \alpha_z & \epsilon_z \gamma_z \end{pmatrix}.
\end{aligned}
\tag{2.19}
$$

The distribution of a Gaussian beam is related to the second moments matrix of the
beam according to

$$
\psi(\mathbf{X}) = \frac{1}{(2\pi)^2 \sqrt{\det \Sigma}} \exp\left(-\frac{1}{2} \mathbf{X}^T \Sigma^{-1} \mathbf{X} \right) = \frac{1}{(2\pi)^2 \epsilon_x \epsilon_z} \exp\left(-\frac{J_x}{\epsilon_x} - \frac{J_z}{\epsilon_z} \right).
\tag{2.20}
$$

2.1.2 Classical $\sigma_z \propto \sqrt{|\eta|}$ Scaling

Now we first reproduce the classical $\sigma_z \propto \sqrt{|\eta|}$ scaling using this longitudinal
Courant-Snyder parameterization. To simplify the discussion further, in this section
and the following we focus on the longitudinal dimension only and the state vector
$\mathbf{X} = (z, \delta)^T$ is used. We treat first the case where there is only one RF placed at a

dispersion-free location. In this case, the linear longitudinal one-turn map observed in the middle of the RF cavity is

$$\mathbf{M} = \begin{pmatrix} 1 & 0 \\ \frac{h}{2} & 1 \end{pmatrix} \begin{pmatrix} 1 & -\eta C_0 \\ 0 & 1 \end{pmatrix} \begin{pmatrix} 1 & 0 \\ \frac{h}{2} & 1 \end{pmatrix} = \begin{pmatrix} 1 - \frac{h}{2}\eta C_0 & -\eta C_0 \\ h - \left(\frac{h}{2}\right)^2 \eta C_0 & 1 - \frac{h}{2}\eta C_0 \end{pmatrix}, \qquad (2.21)$$

with $h = eV_{RF}k_{RF}\cos\phi_s/E_0$ quantifying the RF acceleration gradient, where e is the elementary charge, V_{RF} is the RF voltage, $k_{RF} = 2\pi/\lambda_{RF}$ is the RF wavenumber, ϕ_s is the synchronous phase and $E_0 = \gamma m_e c^2$ is the electron energy. The $R_{56} = -\eta C_0$, a measure for the dependence of z on δ, of the ring and the RF kick h can be viewed as the longitudinal drift space and quadrupole, in correspondence to their transverse counterparts, respectively. Note however that as mentioned before, the R_{56} can be either positive or negative, while the physical length of a drift space is always positive.

The linear stability requires that

$$\left|1 - \frac{h}{2}\eta C_0\right| < 1 \Rightarrow 0 < h\eta C_0 < 4. \qquad (2.22)$$

Actually if the sinousidual modulation waveform is taken into account, the longitudinal dynamics is more accurately modeled by a standard kick map [13]. We then need $h\eta C_0$ be small enough to avoid strong chaotic dynamics. An empirical safe criterion is that $0 < h\eta C_0 \lesssim 0.1$. For rings working in the longitudinal weak focusing regime, $|\nu_s| \ll 1$, we then have

$$1 - \frac{h}{2}\eta C_0 = \cos\Phi_z \approx 1 - \frac{\Phi_z^2}{2} \Rightarrow \Phi_z \approx \begin{cases} -\sqrt{h\eta C_0} & \text{if } \eta > 0, \\ \sqrt{h\eta C_0} & \text{if } \eta < 0. \end{cases} \qquad (2.23)$$

Therefore the longitudinal beta function β_z at the RF center is

$$\beta_{zS} = \frac{\mathbf{M}_{12}}{\sin\Phi_z} \approx \frac{-\eta C_0}{\Phi_z} \approx \sqrt{\frac{\eta C_0}{h}}. \qquad (2.24)$$

In this dissertation, we use the subscript $_S$ to denote results which are the same with that obtained in Sands' classical analysis [5], although the method used here to get these results is different from that of Sands. As $|\nu_s| \ll 1$, therefore

$$\beta_{zS} \gg |-\eta C_0|. \qquad (2.25)$$

We will see later in Sect. 2.1.6 that in a longitudinal strong focusing ring, $|\nu_s|$ can be close to or even larger than 1, and β_z can then be the same level of or smaller than $|-\eta C_0|$.

Using this β_{zS} to represent β_z of the whole ring, we then get the longitudinal emittance obtained in Sands' analysis

$$\epsilon_{zS} = \frac{55}{96\sqrt{3}} \frac{\alpha_F \lambdabar_e^2 \gamma^5 \beta_{zS}}{\alpha_L} \oint \frac{1}{|\rho(s)|^3} ds. \tag{2.26}$$

For a ring consisting of isomagnets, ρ is a positive constant and

$$\alpha_L = J_s \frac{U_0}{2E_0} = J_s \frac{2\pi \lambdabar_e \alpha_F \gamma^3}{3\rho}, \tag{2.27}$$

with $J_s = 2 + \mathcal{D}$ the longitudinal damping partition number [5] and nominally $J_s \approx 2$, we have

$$\beta_{zS} = \frac{-\eta C_0}{\sin \Phi_z} \approx \sqrt{\frac{\eta C_0}{h}},$$

$$\nu_s = \frac{1}{2\pi} \arcsin\left(\frac{-\eta C_0}{\beta_{zS}}\right) \approx -\frac{\eta}{|\eta|} \frac{\sqrt{h\eta C_0}}{2\pi},$$

$$\sigma_{zS} = \sqrt{\epsilon_{zS}\beta_{zS}} \approx \sqrt{\frac{C_q \gamma^2}{J_s \rho}} \sqrt{\frac{\eta C_0}{h}} \approx \sigma_{\delta S}\beta_{zS}, \tag{2.28}$$

$$\sigma_{\delta S} = \sqrt{\epsilon_{zS}\gamma_{zS}} \approx \sqrt{\frac{\epsilon_{zS}}{\beta_{zS}}} \approx \sqrt{\frac{C_q \gamma^2}{J_s \rho}},$$

$$\epsilon_{zS} \approx \frac{C_q \gamma^2}{J_s \rho} \sqrt{\frac{\eta C_0}{h}} \approx \sigma_{zS}\sigma_{\delta S} \approx \sigma_{\delta S}^2 \beta_{zS},$$

where $C_q = \frac{55\lambdabar_e}{32\sqrt{3}} = 3.8319 \times 10^{-13}$ m. Therefore, to generate short bunches in an electron storage ring, we need to implement a quasi-isochronous lattice, i.e., a small η, and a high RF acceleration gradient, i.e., a large h. We also note that the energy spread of an electron beam in the classical analysis is dominantly determined by the beam energy and bending radius of the bending magnets, and has little dependence on the bunch length or global phase slippage of the ring.

2.1.3   Beyond the Classical $\sigma_z \propto \sqrt{|\eta|}$ Scaling

2.1.3.1   Analysis

Using a single β_{zS} to represent that of the whole ring is valid in usual rings where the relative variation of β_z is negligible and the electron distribution in the longitudinal phase space is always upright. But when the global phase slippage is small, the partial phase slippage can be significantly larger than the global one and the variation of β_z and beam orientation in the longitudinal phase space around the ring can be significant, thus the classical $\sigma_z \propto \sqrt{|\eta|}$ scaling fails. Now we present an accurate analysis of this effect using the longitudinal Courant-Snyder formalism.

If there is only a single RF cavity placed at a dispersion-free location in the ring, then at a specific position s_j, the ring can be divided into three parts, with their longitudinal transfer matrices given by

$$\mathbf{T}(s_{RF}, s_j) = \begin{pmatrix} 1 & -\tilde{\eta}(s_{RF}, s_j)C_0 \\ 0 & 1 \end{pmatrix},$$

$$\mathbf{T}(s_{RF}, s_{RF}) = \begin{pmatrix} 1 & 0 \\ h & 1 \end{pmatrix}, \tag{2.29}$$

$$\mathbf{T}(s_j, s_{RF}) = \begin{pmatrix} 1 & -\tilde{\eta}(s_j, s_{RF})C_0 \\ 0 & 1 \end{pmatrix},$$

where $\tilde{\eta}(s_{RF}, s_j) + \tilde{\eta}(s_j, s_{RF}) = \eta$, In the analysis, the RF cavity is assumed to be a zero-length one. The one-turn map at s_j is then

$$\mathbf{M}(s_j) = \mathbf{T}(s_j, s_{RF})\mathbf{T}(s_{RF}, s_{RF})\mathbf{T}(s_{RF}, s_j)$$

$$= \begin{pmatrix} 1 - \tilde{\eta}(s_j, s_{RF})hC_0 & -\eta C_0 + \tilde{\eta}(s_j, s_{RF})\tilde{\eta}(s_{RF}, s_j)hC_0^2 \\ h & 1 - \tilde{\eta}(s_{RF}, s_j)hC_0 \end{pmatrix}. \tag{2.30}$$

Therefore,

$$\beta_z(s_j) = \frac{\mathbf{M}_{12}(s_j)}{\sin \Phi_z} = \frac{-\eta C_0 + \tilde{\eta}(s_j, s_{RF})\tilde{\eta}(s_{RF}, s_j)hC_0^2}{\sin \Phi_z}, \tag{2.31}$$

Note that β_z is always positive, and

$$\frac{d\beta_z(s_j)}{ds_j} = \frac{\left[\tilde{\eta}(s_{RF}, s_j) - \tilde{\eta}(s_j, s_{RF})\right]hC_0}{\sin \Phi_z}\left(\frac{D_x(s_j)}{\rho(s_j)} - \frac{1}{\gamma^2}\right)$$

$$= 2\alpha_z(s_j)\left(\frac{D_x(s_j)}{\rho(s_j)} - \frac{1}{\gamma^2}\right), \tag{2.32}$$

which is different from the conventional relation $\frac{d\beta_{x,y}}{ds} = -2\alpha_{x,y}$ in transverse dimensions [4].

The first term in the numerator of Eq. (2.31) is the conventional global phase slippage. The second term reflects the impact of the partial phase slippage on β_z. In usual rings, the second term is much smaller than the first term, therefore β_z is almost a constant value around the ring. As mentioned, the classical formulas of bunch length $\sigma_z s$, energy spread $\sigma_\delta s$, and longitudinal emittance $\epsilon_z s$ in last section are actually obtained with such approximation. Now with both terms in the numerator of Eq. (2.31) considered, the more accurate formula of the longitudinal emittance is then

$$\epsilon_z = \epsilon_z s\frac{\langle\beta_z\rangle_\rho}{\beta_z s} = \epsilon_z s\left(1 + hC_0\frac{\langle\tilde{\eta}^2(s_j, s_{RF})\rangle_\rho - \eta\langle\tilde{\eta}(s_j, s_{RF})\rangle_\rho}{\eta}\right). \tag{2.33}$$

Note that $\langle\rangle_\rho$ here means the radiation-weighted average around the ring, defined as

$$\langle P \rangle_\rho = \frac{\oint \frac{P}{|\rho(s)|^3} ds}{\oint \frac{1}{|\rho(s)|^3} ds}, \tag{2.34}$$

i.e., the average is actually conducted at places with nonzero bending field. After getting the longitudinal emittance and Courant-Snyder functions, the bunch length and energy spread at a specific location s_i are then

$$\sigma_z(s_i) = \sqrt{\epsilon_z \beta_z(s_i)} \approx \sigma_{zS} \sqrt{\frac{\epsilon_z}{\epsilon_{zS}}} \sqrt{1 - \tilde{\eta}(s_i, s_{\mathrm{RF}}) \tilde{\eta}(s_{\mathrm{RF}}, s_i) \frac{hC_0}{\eta}},$$

$$\sigma_\delta(s_i) = \sqrt{\epsilon_z \gamma_z(s_i)} \approx \sigma_{\delta S} \sqrt{\frac{\epsilon_z}{\epsilon_{zS}}}. \tag{2.35}$$

We remind the readers that the energy spread and γ_z are unchanged outside the RF cavity. In addition, if the contribution of $\frac{1}{\gamma^2}$ is negligible in the definition of η, then α_z and β_z will vary notably only inside the bending magnets. Actually, the chromatic $\mathcal{H}_x$ function, a parameter quantifying the coupling of horizontal emittance to bunch length as can be seen from Eq. (2.19), also changes only inside the bending magnets. Both arguments reveal the fact that in ultrarelativistic cases, bunch length changes only inside the bending magnets. We will see this clearly in Fig. 2.3.

By investigating the bunch length at the RF cavity

$$\sigma_z(s_{\mathrm{RF}}) \approx \sigma_{zS} \sqrt{\frac{\epsilon_z}{\epsilon_{zS}}} = \sigma_{\delta S} \sqrt{\frac{\eta}{hC_0} + \langle \tilde{\eta}^2(s_j, s_{\mathrm{RF}}) \rangle_\rho - \eta \langle \tilde{\eta}(s_j, s_{\mathrm{RF}}) \rangle_\rho C_0}, \tag{2.36}$$

we observe that there exists a lower bunch length limit when η approaches zero

$$\sigma_{z,\mathrm{limit}} = \sigma_{\delta S} \sqrt{\langle \tilde{\eta}^2(s_j, s_{\mathrm{RF}}) \rangle_\rho C_0}. \tag{2.37}$$

This limit is the main consequence of the unavoidable quantum diffusion of longitudinal coordinate in a storage ring. It has little dependence on the global phase slippage and RF voltage, once the beam energy and dispersion function pattern around the ring is given. Since $\sigma_{zS} \propto \sqrt{|\eta|}$, the above bunch length limit means $\frac{\epsilon_z}{\epsilon_{zS}}$ will diverge as η approaches zero. The energy spread will thus diverge in this process.

While the bunch length at the RF cavity will saturate at the limit given by Eq. (2.37) with the decrease of η, the bunch length at other places, from which the partial phase slippage to the RF cavity is large, may first decrease and then increase. The reason is that the increased energy spread will lead to bunch lengthening through the partial phase slippage from the RF cavity to the specific location. In other words, the longitudinal beta function ratio between that at the RF cavity and that at the specific location may increase with the lowering of η.

2.1.3.2 Experimental Verification

The above analysis has been confirmed by numerical simulation as presented in Ref. [2]. Now we introduce our experimental work on this quantum diffusion effect. The experiment was conducted at the Metrology Light Source (MLS) [14–16] of the Physikalisch-Technische Bundesanstalt in Berlin. For usual rings, the bunch length limit given by Eq. (2.37) is a couple of 10 fs to about 100 fs, while the typical bunch length in operation is in 10 ps level. So this effect is negligible in almost all existing rings. However, with the accelerator physics and technologies continue to advance, more ambitious goals of bunch length can be envisioned and realized in the future to benefit more from the electron beam. For example, in an SSMB storage ring, the desired bunch length is sub-micron or even nanometer, which corresponds to sub-fs in unit of time. The quantum diffusion investigated here then becomes the first fundamental issue that needs to be resolved. With such motivation to develop an SSMB light source, and considering that it is a fundamental physical effect by itself, we believe it is important to experimentally verify this effect.

To observe the influence of this effect, we need the second term in the bracket of Eq. (2.33) to be comparable or larger than 1, which is non-trivial for many of the existing storage rings. Other collective and single-particle effects stand in the way before arriving at such a small value of η. However, due to the dedicated quasi-isochronous lattice design and the individually independent magnet power supplies of the MLS storage ring, there is great flexibility in tailoring the lattice optics to obtain a locally large and globally small phase slippage simultaneously, thus opening the possibility to see this effect in an existing machine. Another characteristic making the MLS an ideal test bed of single-particle beam dynamical effects is that it can operate with a beam current ranging from 1 pA (a single electron) to 200 mA.

We have prepared two quasi-isochronous lattice optics at the MLS, named lattice A and B, respectively. Lattice A is the standard quasi-isochronous lattice, while lattice B is developed and dedicated for this experiment. The optical functions of the two lattices are shown in Fig. 2.3. Other related parameters of the two lattices are given in Table 2.1. The key difference of these two lattices is that lattice B has a much larger partial phase slippage and average value of $\langle \beta_z \rangle_\rho$. Therefore, the bunch length limit in lattice B (469 fs at 630 MeV) due to this quantum diffusion is larger than that in lattice A (115 fs at 630 MeV). Note that with the given parameters set, β_z in lattice A is almost a constant value around the ring, while β_z in lattice B varies significantly and at many places is much larger than that in lattice A.

As can be seen in Fig. 2.3, the magnitudes of horizontal dispersion function D_x of lattice B are large at some of the bending magnets, which according to Eq. (2.3) means the local phase slippage increases or decreases sharply within them, leading to a large variation of local phase slippage $\tilde{\eta}$ and β_z. The small global phase slippage η is realized by canceling the contribution of positive and negative D_x at different bending magnets. We remind the readers that this lattice can also be used for the delayed alpha-buckets study in which the momentum differences of particles in different alpha-buckets can be translated into large arrival time differences through the large partial phase slippage [16], which might be useful for some user experiments.

Table 2.1 Parameters of the two lattices of the MLS storage ring used in the experiment

Parameter	Value	Description
C_0	48 m	Ring circumference
E_0	630 MeV	Beam energy
U_0	9.14 keV	Radiation energy loss
f_{RF}	500 MHz	RF frequency
V_{RF}	600 kV	RF voltage
h_{RF}	0.01 m^{-1}	RF acceleration gradient
$\sigma_{\delta S}$	4.4×10^{-4}	Classical energy spread
ϵ_x	197.3 nm	Lattice A
J_s	1.95	Lattice A
$\langle \tilde{\eta}(s_j, s_{RF}) \rangle_\rho$	2.5×10^{-5}	Lattice A
$\sqrt{\langle \tilde{\eta}^2(s_j, s_{RF}) \rangle}_\rho$	1.6×10^{-3}	Lattice A
$\sigma_{z,\text{limit}}$	34 μm (115 fs)	Lattice A
ϵ_x	219.4 nm	Lattice B
J_s	1.95	Lattice B
$\langle \tilde{\eta}(s_j, s_{RF}) \rangle_\rho$	-5.5×10^{-3}	Lattice B
$\sqrt{\langle \tilde{\eta}^2(s_j, s_{RF}) \rangle}_\rho$	6.7×10^{-3}	Lattice B
$\sigma_{z,\text{limit}}$	142 μm (469 fs)	Lattice B

To evaluate the possibility of verifying this effect experimentally, the bunch length and energy spread evolution around the ring in these two lattices have also been presented in Fig. 2.3. Note that the bunch length formula in Eq. (2.35) contains only the contribution from longitudinal emittance. Considering the bunch lengthening by horizontal emittance at dispersive locations, according to Eq. (2.19), the more accurate formula of bunch length is [1, 18]

$$\sigma_z = \sqrt{\epsilon_z \beta_z + \epsilon_x \mathcal{H}_x}. \tag{2.38}$$

Strictly speaking, Courant-Snyder and dispersion functions are only well-defined in a planar uncoupled lattice and only when the RF cavity is placed at a dispersion-free location. For a general coupled lattice, the more accurate SLIM formalism should be referred, i.e.,

$$\sigma_z = \sqrt{2 \sum_{k=I,II,III} \epsilon_k |\mathbf{E}_{k5}|^2},$$

$$\sigma_\delta = \sqrt{2 \sum_{k=I,II,III} \epsilon_k |\mathbf{E}_{k6}|^2}. \tag{2.39}$$

Fig. 2.3 Two lattices used in the experiment. Evolution of **a** $\beta_{x,y}$, **b** D_x and β_z, **c** σ_z, **d** σ_δ, around the ring. In this plot, the RF cavity is placed at $s_{RF} = 0$ m and $V_{RF} = 600$ kV is applied, the global phase slippage used is $\eta = 1 \times 10^{-5}$. The dipole magnets are shown at the top as blue rectangles. Each dipole has a length of 1.2 m and bends the electron trajectory for an angle of $\pi/4$. $\beta_{x,y}$ and D_x are obtained by fitting a model to the BPM-corrector response matrix (LOCO) [17]. The bunch length and energy spread evolution are calculated based on the longitudinal Courant-Snyder formalism and SLIM formalism

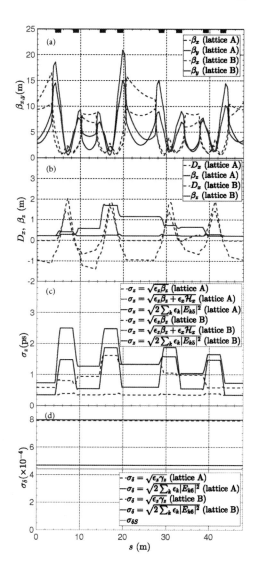

On the other hand, although the RF cavity is placed at a dispersive location in lattice B, we have confirmed that the Courant-Snyder parametrization for beam dynamics analysis in this case is still largely valid, since the difference of result between that given by the longitudinal Courant-Snyder formalism and the more accurate SLIM formalism is very small.

As can be seen in Fig. 2.3, in which the global phase slippage η is lowered to be 1×10^{-5}, which corresponds to a synchrotron frequency of $f_s = 2.2$ kHz with $V_{RF} = 600$ kV, the energy spread grows to be $\sigma_\delta = 7.9 \times 10^{-4}$, while the classical energy spread is $\sigma_{\delta S} = 4.4 \times 10^{-4}$. Such an amount of energy spread growth is detectable by measuring the spectra of Compton-backscattered (CBS) photons from the head-on collision between a CO_2 laser with the electron beam at the MLS [14, 19]. In addition, the bunch length difference in these two lattices are large enough to be observable by evaluating the spectra and power of coherent THz radiation, and invoking streak camera measurement.

To exclude the influence of collective effects, the beam current is lowered to around $6 \mu A$/bunch in a multi-bunch filling mode in the experiment. There is no indication of microwave or other collective instabilities. The beam is stable (no fluctuation of radiation source point observed) and its width and energy spread are independent of the beam current when the single-bunch current is as low as the value applied in the experiment. The horizontal chromaticity has been carefully corrected close to zero (about 0.05) to minimize the beam energy widening arising from the betatron motion of particles as will be reported in Sect. 3.2. The longitudinal chromaticity has also been corrected to a small value to mitigate longitudinal nonlinear dynamics. We note that a large quantum diffusion of longitudinal coordinate (a root-mean-square value of $0.54 \mu m$ or 1.8 fs per turn in lattice B at 630 MeV) actually helps suppress collective beam instability of ultrahigh frequency, as it will disperse any fine time structure in an electron beam like density modulation and energy modulation [7].

To get an idea about the bunch length in the two lattices, first we measure the coherent THz radiation spectra and power as a function of the synchrotron tune in the two lattices. The shorter the electron bunch, the higher frequency range the coherent THz radiation spectra extends and the larger radiation power we can obtain. In the experiment, the synchrotron frequency f_s, thus the global phase slippage η ($f_s \propto \sqrt{|\eta|}$), is controlled by slightly changing the quadrupole currents while keeping the dispersion function pattern unchanged. The THz beamline has its source point at $\frac{\pi}{16}$ bending angle ($s = 38.775$ m) at the 7-th dipole, counted from $s = 0$ m in Fig. 2.3 which is where the RF cavity is placed. To get the coherent synchrotron radiation emission spectra in the THz spectral range, a commercial, Michelson-type FTIR spectrometer (Vertex 80v) in combination with a 4K liquid helium cooled composite silicon bolometer was used for measuring interferograms. After fast Fourier transform of the data, the emitted spectrum can directly be accessed. For this experiment a series of 128 interferograms have been acquired and the average Fourier transformed.

The measured coherent THz radiation power, integrated with wavenumber from 1 to 20 cm^{-1}, together with the theoretical bunch length at the THz observation calculated using Eq. (2.38), are shown in Fig. 2.4a. The measurement results agree with our expectation reasonably well. In particular, we notice that in lattice B, the THz power first increases and then decreases, with the lowering of the synchrotron tune, while the radiation power in lattice A monotonically decreases and then saturates in this process. This observation agrees well with our theoretical prediction of the bunch length evolution in these two lattices. Not presented here, we also notice that the frequency range of the spectra evolves consistently with the integrated power, i.e., a larger THz power corresponds to a higher frequency range coverage. To be more rigorous, we remind the readers that the bunch length in lattice A at the THz

Fig. 2.4 Experiment measurement results and comparison with theory. **a** Theoretical bunch length and measured coherent THz radiation power observed at $s = 38.775$ m, in lattice A and B, respectively. **b** Theoretical, measured raw data (shifted 8 ps downwards) and fitted bunch length at $s = 24$ m, in lattice A and B, respectively. **c** Theoretical and measured electron beam energy spreads σ_δ normalized by the classical energy spread $\sigma_{\delta S}$ versus the synchrotron frequency f_s, in lattice A and B, respectively

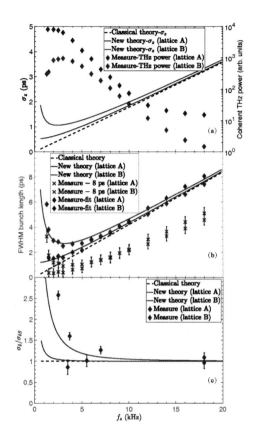

radiation observation point in principle will also diverge, as explained in last section, if we push the phase slippage factor of the ring even closer to zero, which in practice is a demanding work.

During the measurement of coherent THz radiation, we at the same time employed a streak camera to measure the electron bunch length directly. The streak camera at the MLS is installed at the undulator beamline (opposite the RF cavity, $s = 24$ m in Fig. 2.3). For the experiment, the undulator was closed from the "open" gap of 180 mm to 45.7 mm to have the fundamental-mode undulator radiation at a visible wavelength available for the streak camera. The measurement results of bunch length and the comparison with theory is presented in Fig. 2.4b. Note that we have shifted the measured raw data downwards by 8 ps in the plot. The errorbars in the plot are the standard deviation of the fitted results for each single column of the recorded streak camera image. Again we observe the significant difference in the two lattices concerning the bunch length evolution as a function of the synchrotron tune, which agrees qualitatively with theory. However, quantitatively the measured raw data of bunch length deviates notably from the theoretic prediction.

Realizing that there will be unavoidable systematic errors concerning the streak camera measurement because we are close to its resolution limit, we try to use the model below to fit the data with the theory,

$$\Delta z_{\text{fit}} = \sqrt{\Delta z^2_{\text{measure}} - noise^2} - offset, \tag{2.40}$$

where Δz means the bunch length. Note that here we use the full width at half maximum (FWHM), instead of the root mean square, to quantify the bunch length, since in the real case, the bunch profile is unavoidable non-Gaussian to some extent, especially when the phase slippage factor of the ring is small. The *noise* in the above equation is used to model the square sum-type error, while the *offset* accounts for the systematic shift concerning the measurement results. The fitted data (*noise* = 7 ps and *offset* = 3 ps applied) agrees well with the theoretical curve as shown in Fig. 2.4b. We remind the readers that all the data points in the plot are modeled with the same *noise* and *offset*.

Further, we have measured the electron beam energy spread in the two lattices, using the head-on CBS between a CO_2 laser with the electron beam. Note that the RF voltage applied in the above bunch length measurements is 500 kV, while now it is 600 kV when doing the energy spread measurement. The measurement of CBS photon spectra and the evaluation of electron beam energy spread based on it is a well-established method implemented at the MLS, and is used in this experiment to confirm the energy widening as we push the bunch length close to the limit, by lowering the global phase slippage η. More details about this CBS method can be found in Refs. [14, 19]. Quantitative analysis revealing the energy spreads σ_δ normalized by the classical energy spread $\sigma_{\delta S}$, and its comparison with the theoretical prediction from Eq. (2.35) for the two different lattices are shown in Fig. 2.4c. The error bars in Fig. 2.4c are the root-mean-square uncertainties of the measurements and are due to calibration errors and counting statistics. The data acquisition time of a photon spectrum is 15 min. It can be seen from Fig. 2.4 that in lattice B the energy spread grows significantly with the decrease of η, in the figure synchrotron frequency f_s, to the level of 1×10^{-5}, while the energy spread stays almost constant in lattice A. Again the measurement agrees qualitatively with the theory.

There is still some deviation of the measured energy widening and the theoretical prediction for lattice B. Candidate explanations are: first, there is some uncertainty in the determination of synchrotron frequency f_s, especially when f_s is lowered to $2 \sim 3$ kHz, considering the fact that the peak of the synchrotron frequency spectrum then can be as wide as 0.5 kHz; second, there could be some remaining higher-order phase slippages which may contribute to the energy spread growth when η is small due to its impact on the longitudinal phase space bucket, while the theory assumes a linear phase slippage.

The above presented measurements of bunch length and energy spread are very demanding and are moving on the edge of the experimentally accessible parameter space. Nevertheless, we see a nice qualitative agreement with the theory presented in this section, proofing important experimental evidence to support the theoretical analysis. As far as we know, this type of investigation can actually not be performed at any other operating storage ring. However, we recognize that the deviation of the quantitative numbers between the measurements and theory concerning both the bunch length and energy spread emphasizes the need for an even more improved model. Summarizing we state that our experimental work supports the existence of the analyzed quantum diffusion effect, and the argument that the quantum excitation on longitudinal emittance at a given location depends on the longitudinal beta function there. The evidence however is not strong enough to claim this is a fully consistent proof of the effect.

2.1.4 Campbell's Theorem

We point out that quantifying the impact of variation of β_z around the ring on ϵ_z using partial phase slippage variance $\langle \tilde{\eta}^2 \rangle_\rho - \langle \tilde{\eta} \rangle_\rho^2$, as that done in Ref. [6] and also our previous publication Ref. [1], is not generally correct. The reason is that while the photon emission process is stochastic, the evolution of partial phase slippage around the ring is deterministic. So the diffusion of z each turn d_z^2 due to quantum excitation is

$$d_z^2 = \langle z^2 \rangle - \langle z \rangle^2 = C_0^2 \langle \tilde{\eta}^2 \rangle_\rho \langle N \rangle \left\langle \frac{u^2}{E_0^2} \right\rangle, \tag{2.41}$$

instead of

$$d_z^2 = \langle z^2 \rangle - \langle z \rangle^2 = C_0^2 \left(\langle \tilde{\eta}^2 \rangle_\rho - \langle \tilde{\eta} \rangle_\rho^2 \right) \langle N \rangle \left\langle \frac{u^2}{E_0^2} \right\rangle \tag{2.42}$$

as that given in Ref. [6], where $\tilde{\eta}$ is the partial alpha slippage calculated using the final observation location as the ending point, $\langle N \rangle$ is the expected number of emitted photons, u is the photon energy, $\langle u^2 \rangle$ and later also $\langle u \rangle$ mean the average is taken with respect to the photon energy spectrum.

This result can be understood with the help of Campbell's theorem [20]. From this theorem some expectation result for the Poisson point process follows. For example, for the application in synchrotron radiation, we have $\delta = -\sum_i \frac{u_i}{E_0}$, where the subscript $_i$ means the i-th photon emission. Then according to Campbell's theorem we have

$$\langle \delta \rangle = -\langle N \rangle \left\langle \frac{u}{E_0} \right\rangle = -T_{\text{dipole}} \dot{N} \left\langle \frac{u}{E_0} \right\rangle,$$
$$\langle \delta^2 \rangle - \langle \delta \rangle^2 = \langle N \rangle \left\langle \frac{u^2}{E_0^2} \right\rangle = T_{\text{dipole}} \dot{N} \left\langle \frac{u^2}{E_0^2} \right\rangle, \tag{2.43}$$

where $\dot{N}$ is the number of photons emitted per unit time in the dipoles and T_{dipole} is the total time within dipoles. Equation (2.43) is why $\dot{N} \langle u^2 \rangle$ appears so often in the calculation of energy spread, emittance, etc., in electron storage ring physics. Note that the relation in Eq. (2.43) holds as long as the radiation is a Poisson point process. It is independent of whether $\langle u \rangle = 0$ or not, and is also independent of the detailed spectrum of the photon energy. In other words, the key of a Poisson point process is the randomness in whether there is a kick or not, i.e, the kick number, and not in the randomness of the size of the kicks. The importance of this theorem for electron dynamics was first pointed out by Sands [21]. A proof can be found in the article of Rice [22] and a less rigorous but simpler one in the lecture note of Jowett [23].

Now we can understand Eq. (2.41) as follows. Suppose that the RF is our observation point. We divide the ring into many sections, and in each section $\tilde{\eta}(s_{RF}, s_j)$ does not change much. Then the change of electron longitudinal coordinate in one turn is $z = \sum_j z_j$, with $z_j = \sum_i C_0 \tilde{\eta}(s_{RF}, s_{ji}) \frac{u_{ji}}{E_0}$ the contribution due to photon emissions within the section j. According to Campbell's theorem, the variance of z_j is

$$\text{Var}(z_j) = C_0^2 \tilde{\eta}^2(s_{RF}, s_j) t_j \dot{N} \left\langle \frac{u^2}{E_0^2} \right\rangle, \tag{2.44}$$

where t_j is the time within the dipoles in section j. As the photon emissions in different sections are uncorrelated, then the variance of z is the sum of variance of z_j

$$\langle z^2 \rangle - \langle z \rangle^2 = C_0^2 \frac{\sum_j \left[\tilde{\eta}^2(s_{RF}, s_j) t_j \right]}{T_{total}} T_{total} \dot{N} \left\langle \frac{u^2}{E_0^2} \right\rangle$$
$$= C_0^2 \langle \tilde{\eta}^2(s_{RF}, s_j) \rangle_\rho \langle N \rangle \left\langle \frac{u^2}{E_0^2} \right\rangle, \tag{2.45}$$

in which $T_{total} = \sum_j t_j$ is the total time within the dipoles. So now we have obtained Eq. (2.41) following Campbell's theorem.

We can also view the above argument from another way. Given the same dispersion function pattern, which means the same $\langle \tilde{\eta}^2 \rangle_\rho - \langle \tilde{\eta} \rangle_\rho^2$ as it is independent of the observation point, a different longitudinal beta function pattern can be generated if the RF is placed at a different location, therefore resulting in a different longitudinal emittance according to Eq. (2.14).

Changing the RF location means shifting $\tilde{\eta}(s_j, s_{RF})$ up or down as a whole. According to Eq. (2.33), the equilibrium emittance is a parabolic function of the shifted value. When the RF is placed at a location such that $\langle \tilde{\eta}(s_j, s_{RF}) \rangle_\rho = \frac{\eta}{2}$, we arrive at the minimum longitudinal emittance

$$\epsilon_{z,min} = \epsilon_{zS} \left(1 + \frac{\langle \tilde{\eta}^2 \rangle_\rho - \langle \tilde{\eta} \rangle_\rho^2 - \left(\frac{\eta}{2} \right)^2}{\eta C_0 / h} C_0^2 \right). \tag{2.46}$$

The maximum longitudinal emittance is realized when the RF is placed at a place such that $\left| \langle \tilde{\eta}(s_j, s_{RF}) \rangle_\rho - \frac{\eta}{2} \right|$ reaches the maximum possible value. When the minimum longitudinal emittance is reached, the bunch length at the RF is

$$\sigma_{z,\min}(s_{\mathrm{RF}}) = \sigma_{\delta\mathrm{S}}\sqrt{\frac{\eta C_0}{h} + \left[\langle\tilde{\eta}^2\rangle_\rho - \langle\tilde{\eta}\rangle_\rho^2 - \left(\frac{\eta}{2}\right)^2\right]C_0^2}. \qquad (2.47)$$

In the case of ultrasmall η, we have

$$\sigma_{z,\min}(s_{\mathrm{RF}}) \approx \sigma_{\delta\mathrm{S}}\sqrt{\langle\tilde{\eta}^2\rangle_\rho - \langle\tilde{\eta}\rangle_\rho^2}\,C_0, \qquad (2.48)$$

and

$$\epsilon_{z,\min} \approx \frac{\epsilon_z\mathrm{S}}{\sqrt{\eta C_0/h}}\left(\sqrt{\eta C_0/h} + \frac{\left(\langle\tilde{\eta}^2\rangle_\rho - \langle\tilde{\eta}\rangle_\rho^2\right)C_0^2}{\sqrt{\eta C_0/h}}\right)$$

$$\geq 2\sigma_{\delta\mathrm{S}}^2\sqrt{\langle\tilde{\eta}^2\rangle_\rho - \langle\tilde{\eta}\rangle_\rho^2}\,C_0. \qquad (2.49)$$

The equality holds when $\beta_{z\mathrm{S}} = \sqrt{\frac{\eta C_0}{h}} = C_0\sqrt{\langle\tilde{\eta}^2\rangle_\rho - \langle\tilde{\eta}\rangle_\rho^2}$. Therefore, the variance of partial phase slippage can be viewed as a parameter to quantify the lowest possible contribution of this effect to the equilibrium bunch length at the RF and the longitudinal emittance with a dispersion function pattern given, if we can choose the location of the RF as we want. However, in a real machine, the RF location is fixed, and Eqs. (2.33) and (2.35) should be referred. This is why we state that using $\langle\tilde{\eta}^2\rangle_\rho - \langle\tilde{\eta}\rangle_\rho^2$ to quantify the impact of this effect is not generally correct.

2.1.5 Minimizing Longitudinal Emittance

It is clear that the quantum diffusion of z needs to be carefully treated for the realization and long-term maintenance of ultrashort bunch or small longitudinal emittance in either a multi-pass device or a single-pass transport line with bending magnets and large dispersion. A lower operating energy is preferred for suppressing the strength of quantum excitation. Note that the energy scaling laws of this effect are different in the one-turn or single-pass and steady-state cases; for the single-pass case, i.e., Eq. (2.41), the root-mean-square diffusion of longitudinal coordinate $d_z \propto \gamma^{2.5}$, while for the steady-state case, i.e., Eq. (2.48), $\sigma_{z,\min} \propto \gamma$, because the radiation damping time also depends on γ.

As can be seen in Eq. (2.14), the longitudinal beta function β_z with respect to the longitudinal dimension plays a role similar to that of the chromatic function $\mathcal{H}_x$ in the transverse dimension. As both the longitudinal and transverse emittances originate from quantum excitation, for a ring consisting of identical isochronous cells, the same scaling law of the theoretical minimum emittance (TME), i.e., $\epsilon_{x,z,\mathrm{TME}} \propto \gamma^2\theta^3$, concerning the beam energy and bending angle of the dipole can be expected. Note that the TME is independent of the bending radius. But we will show soon that the bunch length limit does depend on the bending radius. According to the scaling, a ring consisting of a larger number of isochronous cells, each with a smaller bending

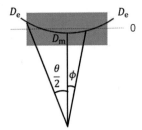

Fig. 2.5 A symmetric dispersion function pattern which makes each half of the bending magnet isochronous, which is desired in minimizing the theoretical minimum bunch length and longitudinal emittance when there is only a single RF or laser modulator in the ring. The lattice design for the realization of the dispersion function pattern can be found in Ref. [25]

angle, can better minimize the emittance than a ring consisting of fewer cells with larger bending. Generally, it is easier to realize small emittance in a larger ring.

Equation (2.48) gives the lower bunch length limit by optimizing the location of RF cavity, with a given dispersion function pattern. To make this limit as small as possible, in addition to ensuring a small global phase slippage, the variation in the partial phase slippage should also be well confined by means of dedicated lattice design. More specifically, the strategy is to tailor the horizontal dispersion function, thus to minimize the longitudinal beta function at the bending magnets. At the MLS, the small global phase slippage is achieved by means of an overall integration cancellation between the large positive and large negative horizontal dispersions at different dipoles [15, 16]. Therefore, the partial phase slippage varies sharply within the dipoles, leading to a large partial phase slippage variation and significant quantum diffusion of z. To obtain small global and partial phase slippages simultaneously, such cancellation should be done as locally as possible, and the magnitudes of the dispersion at the dipoles should also be minimized, thus making the partial phase slippage vary as gently as possible. In other words, each partial component of the ring should be made as isochronous as possible. In this sense, the dispersion function pattern in Fig. 2.5 is the most locally isochronous bending magnet [24], i.e., each half of the bending magnet is isochornous.

2.1.5.1 Constant Bending Radius

Now we present some quantitative analysis of the minimization of longitudinal emittance. In this section we use the partial R_{56}, defined as

$$F(s_2, s_1) \equiv -\tilde{\eta}(s_2, s_1)C_0 = -\int_{s_1}^{s_2} \left(\frac{D_x(s)}{\rho(s)} - \frac{1}{\gamma^2} \right) ds, \qquad (2.50)$$

for the analysis. As can be seen Eqs. (2.48) and (2.49), it is $\sqrt{\langle F^2 \rangle_\rho - \langle F \rangle_\rho^2}$ that determines the theoretical minimum bunch length and longitudinal emittance. Here we present an analysis of the above scaling by evaluating $\sqrt{\langle F^2 \rangle_\rho - \langle F \rangle_\rho^2}$ of a single isochronous bending magnet with constant bending-radius, and whose dispersion function is symmetric with respect to the dipole middle point as shown in Fig. 2.5. To simplify the analysis, we start from the middle point of the dipole where the dispersion angle is zero $D_m' = 0$, then the dispersion as a function of angle ϕ is

$$D(\phi) = D_m \cos \phi + \rho(1 - \cos \phi), \tag{2.51}$$

where D_m is the dispersion at the middle of the dipole. As we are mainly interested in the relativistic case, ignoring the contribution of $\frac{1}{\gamma^2}$ on F, the condition of isochronicity of each half of the dipole is

$$\int_0^{\frac{\theta}{2}} D(\phi) d\phi = 0, \tag{2.52}$$

with θ the bending angle of each dipole. Substituting Eq. (2.51) into Eq. (2.52), we get

$$D_m = \rho \left(1 - \frac{\frac{\theta}{2}}{\sin \frac{\theta}{2}} \right) \approx -\frac{1}{24} \rho \theta^2,$$

$$D_e = \rho \left(1 - \frac{\frac{\theta}{2}}{\tan \frac{\theta}{2}} \right) \approx \frac{1}{12} \rho \theta^2, \tag{2.53}$$

where D_e is the dispersion at the entrance and exit of the dipole. Then we have

$$F(\phi) = \int_0^\phi D(\beta) d\beta = \rho \left(\phi - \frac{\frac{\theta}{2}}{\sin \frac{\theta}{2}} \sin \phi \right),$$

$$\langle F \rangle_\rho = \frac{1}{\theta} \int_{-\frac{\theta}{2}}^{\frac{\theta}{2}} F(\phi) d\phi = 0,$$

$$\langle F^2 \rangle_\rho = \frac{1}{\theta} \int_{-\frac{\theta}{2}}^{\frac{\theta}{2}} F^2(\phi) d\phi = \frac{1}{6} \rho^2 \left[2 \left(-6 + \left(\frac{\theta}{2} \right)^2 \right) + 9 \frac{\frac{\theta}{2}}{\tan \frac{\theta}{2}} + 3 \frac{\left(\frac{\theta}{2} \right)^2}{\sin \left(\frac{\theta}{2} \right)^2} \right],$$

$$\sqrt{\langle F^2 \rangle_\rho - \langle F \rangle_\rho^2} \approx \frac{\sqrt{210}}{2520} \rho \theta^3. \tag{2.54}$$

Note that the F in this section is defined with the middle point of the bending magnet as the starting point. Therefore, for a ring consisting of such isochronous isomagnets (note that the global phase slippage of the ring is non-zero for a stable beam motion), we have

$$\sigma_{z,\min} \approx \sigma_{\delta S}\sqrt{\langle F^2 \rangle_\rho - \langle F \rangle_\rho^2} = \frac{\sqrt{210}}{2520}\sqrt{\frac{C_q}{J_s}}\sqrt{\rho}\gamma\theta^3 \propto \sqrt{\rho}\gamma\theta^3, \tag{2.55}$$

and

$$\epsilon_{z,\min} \approx 2\sigma_{\delta S}^2\sqrt{\langle F^2 \rangle_\rho - \langle F \rangle_\rho^2} = \frac{\sqrt{210}}{1260}\frac{C_q}{J_s}\gamma^2\theta^3 \propto \gamma^2\theta^3. \tag{2.56}$$

We can also derive the above emittance scaling using directly the longitudinal beta function evolution in the dipole

$$\beta_z(\phi) = \beta_{zm} - 2\alpha_{zm}F(\phi) + \gamma_{zm}F^2(\phi), \tag{2.57}$$

where α_{zm}, β_{zm}, γ_{zm} are the longitudinal Courant-Snyder functions at the middle of the dipole. Then

$$\begin{aligned}
\langle \beta_z \rangle_\rho &= \frac{1}{\theta}\int_{-\frac{\theta}{2}}^{\frac{\theta}{2}} \beta_z(\phi)d\phi \\
&= \frac{\rho^2\left(\alpha_{zm}^2 + 1\right)\left[2\left(\theta^2 - 24\right) + 3\theta(\theta + 3\sin(\theta))\csc^2\left(\frac{\theta}{2}\right)\right]}{24\beta_{zm}} + \beta_{zm} \\
&\approx \frac{\rho^2\theta^6\left(\alpha_{zm}^2 + 1\right)}{30240\beta_{zm}} + \beta_{zm}.
\end{aligned} \tag{2.58}$$

The minimum average of β_z, $\langle \beta_z \rangle_{\rho,\min}$, thus the minimum longitudinal emittance, $\epsilon_{z,\min}$, is realize when

$$\alpha_{zm} = 0, \quad \beta_{zm} = \frac{\langle \beta_z \rangle_{\rho,\min}}{2} = \frac{\sqrt{210}}{2520}\rho\theta^3. \tag{2.59}$$

The corresponding minimum longitudinal emittance is

$$\epsilon_{z,\min} = \frac{55}{96\sqrt{3}}\frac{\alpha_F \lambda_e^2 \gamma^5}{\alpha_L}\frac{2\pi\langle \beta_z \rangle_{\rho,\min}}{\rho^2} = \frac{\sqrt{210}}{1260}\frac{C_q}{J_s}\gamma^2\theta^3, \tag{2.60}$$

which is the same as that given in Eq. (2.56). If the longitudinal damping partition number $J_s = 2$, then for practical use we have

$$\begin{aligned}
\sigma_{z,\min}[\mu m] &\approx 4.93\rho^{\frac{1}{2}}[m]E_0[\text{GeV}]\theta^3[\text{rad}], \\
\epsilon_{z,\min}[nm] &\approx 8.44E_0^2[\text{GeV}]\theta^3[\text{rad}].
\end{aligned} \tag{2.61}$$

This is the main result of our analysis of theoretical minimum bunch length and longitudinal emittance in a longitudinal weak focusing ring. A comprehensive study of minimizing longitudinal emittance can also be found in Ref. [4].

Actually if we relax the condition of isochronousity for each dipole, and make D_m and D'_m also variables that can be optimized, then we have

$$
\langle \beta_z \rangle_\rho = \frac{1}{\theta} \int_{-\frac{\theta}{2}}^{\frac{\theta}{2}} \beta_z(\phi) d\phi
$$

$$
\approx \beta_{zm} + \frac{D_m^2 \theta^2}{12\beta_{zm}} + \frac{4D_m \rho \theta^4}{240\beta_{zm}} + \frac{10\rho^2\theta^6 + 504 D_m'^2\rho^2\theta^4 - 15 D_m'^2\rho^2\theta^6}{161280\beta_{zm}}.
$$

$$(2.62)$$

When $D_m = -\frac{\rho\theta^2}{40}$, $D'_m = 0$, and $\beta_{zm} = \frac{\sqrt{7}}{840}\rho\theta^3$, we reach the minimal of $\langle \beta_z \rangle_{\rho,\min} = \frac{\sqrt{7}}{420}\rho\theta^3$ and longitudinal emittance

$$
\epsilon_{z,\min} = \frac{\sqrt{7}}{420}\frac{C_q}{J_s}\gamma^2\theta^3.
$$

$$(2.63)$$

This is also the result Eq. (29) in Ref. [4]. The emittance given in Eq. (2.63) is smaller than that given in Eq. (2.60) or equivalently Eq. (2.61). However, we should note that when isochronousity of each dipole is broken, to make the longitudinal beta function in different dipoles identical, we need RF kick between each two neighboring dipoles to adjust α_z there, or we need a very long drift space between them if we consider the contribution of $\frac{1}{\gamma^2}$ on R_{56}. This means at least N RFs or laser modulators are needed if there are N dipoles in the ring. This is not very feasible in reality. In addition, placing RFs at dispersive locations will make the dynamics becomes transverse-longitudinal coupled. In the more-confronted or practical case of a single RF in the ring, Eq. (2.61) is a more self-consistent evaluation of the theoretical minimum bunch length and longitudinal emittance.

Since high-power EUV radiation is of particular interest for EUV lithography [26], let us now do some evaluation based on our investigations to see if we can realize high-power EUV radiation in a longitudinal weak focusing SSMB storage ring. For coherent 13.5 nm EUV radiation generation, we need an electron bunch length around 3 nm or shorter. The lower limit of bunch length $\sigma_{z,\min}$ should be smaller than this desired bunch length to avoid significant energy widening. Here we assume $\sigma_{z,\min} \leq$ 2 nm. If $E_0 = 400$ MeV and $\rho = 4$ m ($B = 0.334$ T), then $\sigma_{\delta S} = 1.7 \times 10^{-4}$. To realize $\sigma_{z,\min} \leq 2$ nm, according to Eq. (2.61), we need $\theta \leq 0.0797$ rad $\approx \frac{2\pi \text{ rad}}{79}$, which means 79 bending magnets are required in the ring. If the length of each isochronous cell with a single bending magnet can be designed to be around 2 m, then the circumference of the ring can be about 180 m, considering the sections of laser modulation, radiation generation and the energy supply system, etc.

Applying the lowest phase slippage factor realizable in practice at present, which is about $\eta = 1 \times 10^{-6}$, to realize a 3 nm bunch length in such a ring, the required energy chirp strength is then $h = \eta C_0 \left(\frac{\sigma_{\delta S}}{\sigma_{zS}}\right)^2 = 5.88 \times 10^5$ m^{-1}. The effective modulation voltage of a laser modulator using a planar undulator is related to the laser and undulator parameters as [27]

$$V_L = \frac{[JJ]K}{\gamma} \sqrt{\frac{4P_L Z_0 Z_R}{\lambda_L}} \tan^{-1}\left(\frac{L_u}{2Z_R}\right), \qquad (2.64)$$

in which $[JJ] = J_0(\chi) - J_1(\chi)$ with J_n the n-th order Bessel function of the first kind and $\chi = \frac{K^2}{4+2K^2}$, $K = \frac{eB_0}{m_e c k_u} = 0.934 \cdot B_0[T] \cdot \lambda_u[cm]$ is the dimensionless undulator parameter, determined by the undulator period and magnetic flux density, P_L is the modulation laser power, $Z_0 = 376.73\ \Omega$ is the impedance of free space, Z_R is the Rayleigh length of the laser, L_u is the undulator length. The linear energy chirp strength around zero-crossing phase is therefore

$$h = \frac{eV_L}{E_0}k_L = \frac{e[JJ]K}{\gamma^2 mc^2}\sqrt{\frac{4P_L Z_0 Z_R}{\lambda_L}} \tan^{-1}\left(\frac{L_u}{2Z_R}\right)k_L, \qquad (2.65)$$

where $k_L = 2\pi/\lambda_L$ is the wavenumber of the modulation laser. If $\lambda_L = 270$ nm, $\lambda_u = 4$ cm, $B_0 = 1.02$ T, $L_u = 1.6$ m, $Z_R = \frac{L_u}{3}$, then the required modulation laser power to get $h = 5.88 \times 10^5$ m^{-1} is $P_L = 2.75$ GW. At present, 1 MW stored average laser power is the state-of-art value we can realize in practice for an optical enhancement cavity. So we can only operate the cavity in pulsed mode, which means the average radiation power will be limited. To put it another way, if a CW optical cavity and a practical global phase slippage is applied, a longitudinal weak focusing SSMB storage ring can only realize bunch length as low as tens of nanometer with a beam energy of several hundred MeV. Such an SSMB ring can provide high-power radiation with wavelength $\lambda_R \gtrsim 100$ nm. We have presented in Table 6.1 of the final chapter an example parameters set of SSMB ring for high-power infrared radiation. We remind the readers again that considering the nonlinear modulation waveform, we actually need $0 < h\eta C_0 \lesssim 0.1$ to avoid strong chaotic dynamics in a longitudinal weak focusing ring.

The above limitation of longitudinal weak focusing scheme is the motivation for us to develop the longitudinal strong focusing SSMB and transverse-longitudinal coupling SSMB, or generalized longitudinal strong focusing SSMB, to compress the bunch length further for coherent EUV and soft X-ray radiation generation. We will present the details of these advanced scenarios in the following part of this dissertation.

2.1.5.2 Transverse Gradient Bends

The above analysis of theoretical minimum bunch length and emittance is for a constant bending radius. To minimize the longitudinal emittance further, transverse and longitudinal gradient bending magnets (TGB and LGB) can be invoked. Below we conduct some calculations based on the similar dispersion configuration as shown in Fig. 2.5, but this time using a TGB. The Hill's equation for the dispersion is

$$\frac{d^2 D(s)}{ds^2} + \left(\frac{1}{\rho(s)^2} - k(s)\right) D(s) = \frac{1}{\rho(s)}. \tag{2.66}$$

For simplicity, here we only investigate the case of a constant bending radius $\rho(s) = \rho$ and a constant transverse gradient $k(s) = k$. To simplify the writing, we denote

$$g \equiv \frac{1}{\rho^2} - k. \tag{2.67}$$

If $g > 0$, then the solution of Eq. (2.66) is

$$D(s) = D_i \cos\left(\sqrt{g}s\right) + D_i' \frac{\sin\left(\sqrt{g}s\right)}{\sqrt{g}} + \frac{1}{g\rho}\left[1 - \cos\left(\sqrt{g}s\right)\right],$$

$$D'(s) = -D_i\sqrt{g}\sin\left(\sqrt{g}s\right) + D_i'\cos\left(\sqrt{g}s\right) + \frac{1}{\sqrt{g}\rho}\sin\left(\sqrt{g}s\right), \tag{2.68}$$

where D_i and D_i' are the initial dispersion and dispersion angle at the origin $s = 0$ m, respectively. If $g < 0$, then the solution of Eq. (2.66) is

$$D(s) = D_i \cosh\left(\sqrt{|g|}s\right) + D_i' \frac{\sinh\left(\sqrt{|g|}s\right)}{\sqrt{|g|}} + \frac{1}{|g|\rho}\left[-1 + \cosh\left(\sqrt{|g|}s\right)\right],$$

$$D'(s) = -D_i\sqrt{|g|}\sinh\left(\sqrt{|g|}s\right) + D_i'\cosh\left(\sqrt{|g|}s\right) + \frac{1}{\sqrt{|g|}\rho}\sinh\left(\sqrt{|g|}s\right). \tag{2.69}$$

Below, we present the derivations for the case of $g > 0$ and the results are similar when $g < 0$. Like our previous calculations, we set the origin at the middle of the dipole where $D_m' = 0$, the dispersion as a function of angle ϕ is then

$$D(\phi) = D_m \cos\left(\sqrt{g}\rho\phi\right) + \frac{1}{g\rho}\left[1 - \cos\left(\sqrt{g}\rho\phi\right)\right]. \tag{2.70}$$

Substitute Eq. (2.70) into the isochronicity condition Eq. (2.52), we get

$$D_m = \frac{1}{g\rho}\left[1 - \frac{\sqrt{g}\rho\frac{\theta}{2}}{\sin\left(\sqrt{g}\rho\frac{\theta}{2}\right)}\right] \approx -\frac{1}{24}\rho\theta^2\left(1 + \frac{7}{240}g\rho^2\theta^2\right),$$

$$D_e = D\left(\frac{\theta}{2}\right) = \frac{1}{g\rho}\left[1 - \frac{\sqrt{g}\rho\frac{\theta}{2}}{\tan\left(\sqrt{g}\rho\frac{\theta}{2}\right)}\right] \approx \frac{1}{12}\rho\theta^2\left(1 + \frac{1}{60}g\rho^2\theta^2\right). \tag{2.71}$$

The $\sqrt{\langle F^2 \rangle_\rho - \langle F \rangle_\rho^2}$ in this case can be calculated as follows

$$F(\phi) = \int_0^\phi D(\phi')d\phi' = \frac{1}{g\rho}\left[\phi - \frac{\frac{\theta}{2}}{\sin\left(\sqrt{g}\rho\frac{\theta}{2}\right)}\sin\left(\sqrt{g}\rho\phi\right)\right],$$

$$\langle F \rangle_\rho = \frac{1}{\theta}\int_{-\frac{\theta}{2}}^{\frac{\theta}{2}} F(\phi)d\phi = 0$$

$$\langle F^2 \rangle_\rho = \frac{1}{\theta}\int_{-\frac{\theta}{2}}^{\frac{\theta}{2}} F^2(\phi)d\phi$$

$$= \frac{1}{6g^3\rho^4}\left[-12 + 2g\rho^2\left(\frac{\theta}{2}\right)^2 + 9\frac{\sqrt{g}\rho\left(\frac{\theta}{2}\right)}{\tan\left(\sqrt{g}\rho\left(\frac{\theta}{2}\right)\right)} + 3\frac{\left(\sqrt{g}\rho\left(\frac{\theta}{2}\right)\right)^2}{\sin\left(\sqrt{g}\rho\left(\frac{\theta}{2}\right)\right)^2}\right],$$

$$\sqrt{\langle F^2 \rangle_\rho - \langle F \rangle_\rho^2} \approx \frac{\sqrt{210}}{2520}\rho\theta^3\left(1 + \frac{g\rho^2\theta^2}{40}\right).$$

$$\tag{2.72}$$

For example, to reduce $\sqrt{\langle F^2 \rangle_\rho - \langle F \rangle_\rho^2}$ by a factor of two compared to the case of no transverse gradient, we need

$$\frac{g\rho^2\theta^2}{40} = -\frac{1}{2} \Rightarrow g = -\frac{20}{(\rho\theta)^2}. \tag{2.73}$$

For the example shown in last section, $\rho = 4$ m, $\theta = \frac{2\pi}{79}$, then $k = 63$ m^{-2}, which is a practical gradient.

Besides the influence on $\sqrt{\langle F^2 \rangle_\rho - \langle F \rangle_\rho^2}$, the transverse gradient may also affect the damping partition and hence has an impact on the bunch length and longitudinal emittance. For the specific case of a constant bending radius with a constant transverse gradient we are treating, we have

$$I_2 = \oint \frac{1}{\rho^2}ds = \frac{2\pi}{\rho},$$

$$I_4 = \oint \frac{D_x}{\rho^3}(1 + 2\rho^2 k)ds = 0, \tag{2.74}$$

where I_2 and I_4 are the radiation integrals [5]. Then $J_s = 2 + \frac{I_4}{I_2} = 2$. So a dipole with a constant bending radius and a constant transverse gradient is not very flexible in controlling the damping partition number, due to the constraint of isochronous condition. A varying transverse gradient may be helpful to minimize the longitudinal emittance, and optimization of the transverse gradient profile based on numerical method can be invoked. The application of TGB can also be analyzed following the same formalism, which we do not detail in this dissertation.

2.1.5.3 Transverse Emittance Scaling

For completeness, now we present the horizontal emittance scaling in a longitudinal weak focusing SSMB storage ring. To realize the dispersion function pattern shown in Fig. 2.5, in thin-lens approximation, the horizontal optical functions at the dipole middle point are correspondingly[1]

$$\beta_{xm} = -\frac{\rho\theta}{3}\frac{\sin \Phi_x}{1 + \cos \Phi_x}, \quad \alpha_{xm} = 0, \tag{2.75}$$

with Φ_x the betatron phase advance per isochronous cell, which usually lies in $(\pi, 2\pi)$. We have assumed there is only a single dipole each isochronous cell. The normalized eigenvector corresponding to the horizontal plane at the dipole middle point is

$$\mathbf{E}_I(0) = \frac{1}{\sqrt{2}}\begin{pmatrix} \sqrt{\beta_{xm}} \\ \frac{i}{\sqrt{\beta_{xm}}} \\ 0 \\ 0 \\ \frac{i}{\sqrt{\beta_{xm}}}D_m \\ 0 \end{pmatrix} e^{i\Phi_I}. \tag{2.76}$$

The transfer matrix of a sector dipole with no transverse gradient is

$$\mathbf{S}(\alpha) = \begin{pmatrix} \cos \alpha & \rho \sin \alpha & 0 & 0 & 0 & \rho(1 - \cos \alpha) \\ -\frac{\sin \alpha}{\rho} & \cos \alpha & 0 & 0 & 0 & \sin \alpha \\ 0 & 0 & 1 & \rho\alpha & 0 & 0 \\ 0 & 0 & 0 & 1 & 0 & 0 \\ -\sin \alpha & -\rho(1 - \cos \alpha) & 0 & 0 & 1 & \rho\left(\frac{\alpha}{\gamma^2} + \alpha - \sin \alpha\right) \\ 0 & 0 & 0 & 0 & 0 & 1 \end{pmatrix}. \tag{2.77}$$

Then

$$\mathcal{H}_x(\alpha) = 2|\mathbf{E}_{I5}(\alpha)|^2 = 2|\mathbf{S}(\alpha)\mathbf{E}_{I5}(0)|^2$$

$$= -\rho\left\{\frac{\theta}{3}\frac{\sin \Phi_x}{1 + \cos \Phi_x}\sin^2 \alpha + \frac{3}{\theta}\frac{1 + \cos \Phi_x}{\sin \Phi_x}\left[\frac{\theta^2}{24} + (1 - \cos \alpha)\right]^2\right\}. \tag{2.78}$$

Note that $\mathcal{H}_x(-\alpha) = \mathcal{H}_x(\alpha)$. Putting in $\alpha_H \approx \frac{\alpha_L}{2} \approx \frac{U_0}{2E_0} = \frac{1}{2}C_\gamma\frac{E_0^3}{\rho}$, then according to Eq. (2.14), the equilibrium horizontal emittance is

---

[1] Private communication with Zhilong Pan.

$$\epsilon_x = \frac{55}{96\sqrt{3}} \frac{\alpha_F \lambda_e^2 \gamma^5}{\frac{1}{2} C_\gamma \frac{E_0^3}{\rho}} \frac{2\pi}{\theta} \times 2 \times \int_0^{\frac{\theta}{2}} \frac{\mathcal{H}_x(\alpha)}{\rho^2} d\alpha$$

$$\approx -2\pi \frac{55}{24\sqrt{3}} \frac{\alpha_F \lambda_e^2 \gamma^5}{C_\gamma E_0^3} \left[\frac{1}{72} \tan\left(\frac{\Phi_x}{2}\right) + \frac{1}{80} \cot\left(\frac{\Phi_x}{2}\right) \right] \theta^3. \tag{2.79}$$

Putting in the numbers, we have

$$\epsilon_x[\text{nm}] = -366.5 E_0^2[\text{GeV}] \theta^3[\text{rad}] \left[\frac{1}{9} \tan\left(\frac{\Phi_x}{2}\right) + \frac{1}{10} \cot\left(\frac{\Phi_x}{2}\right) \right]. \tag{2.80}$$

Note that the horizontal emittance diverges as Φ_x approaches π or 2π.

2.1.6 Longitudinal Strong Focusing

2.1.6.1 Analysis

The analysis in the above sections considers the case with only a single RF. When there are multiple RFs, for the longitudinal dynamics, it is similar to implement multiple quadrupoles in the transverse dimension, and the beam dynamics can have more possibilities. Longitudinal strong focusing scheme for example can be invoked [8, 28], not unlike its transverse counterpart which is the foundation of modern high-energy accelerators [29, 30]. The linear beam dynamics with multiple RFs can be treated using SLIM the same way as that with a single RF. When all the RFs are placed at dispersion-free locations, the Courant-Snyder parametrization can be applied as analyzed in previous sections. Here we use a setup with two RFs as an example to show the scheme of manipulating β_z around the ring. The schematic layout of the ring is shown in Fig. 2.6. The treatment of cases with more RFs is similar.

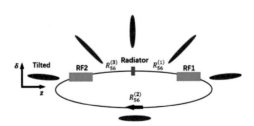

Fig. 2.6 A schematic layout of a storage ring using two RF systems for longitudinal strong focusing and an example beam distribution evolution in the longitudinal phase space. Note that the tilted angles of the beam distribution and bunch length ratios at different places do not strictly correspond to the parameters in Table 2.2, but only to present the qualitative characteristics

We divide the ring into five sections, i.e., three longitudinal drifts (R_{56}) and two longitudinal quadrupole kicks (h), with the transfer matrices given by

$$\mathbf{T}_{D1} = \begin{pmatrix} 1 & R_{56}^{(1)} \\ 0 & 1 \end{pmatrix}, \quad \mathbf{T}_{RF1} = \begin{pmatrix} 1 & 0 \\ h_1 & 1 \end{pmatrix}, \quad \mathbf{T}_{D2} = \begin{pmatrix} 1 & R_{56}^{(2)} \\ 0 & 1 \end{pmatrix},$$

$$\mathbf{T}_{RF2} = \begin{pmatrix} 1 & 0 \\ h_2 & 1 \end{pmatrix}, \quad \mathbf{T}_{D3} = \begin{pmatrix} 1 & R_{56}^{(3)} \\ 0 & 1 \end{pmatrix}. \tag{2.81}$$

Then the one-turn map at the radiator center is

$$\mathbf{M}_R = \mathbf{T}_{D3}\mathbf{T}_{RF2}\mathbf{T}_{D2}\mathbf{T}_{RF1}\mathbf{T}_{D1}. \tag{2.82}$$

Linear stability requires that $|\mathrm{Tr}\,(\mathbf{M}_R)| < 2$, where $\mathrm{Tr}()$ means the trace of. For the generation of coherent radiation, we usually want the bunch length to reach its minimum at the radiator, then we need $\alpha_z = 0$ for $\mathbf{M}_R$.

With the primary purpose to present the principle, instead of a detailed design, here for simplicity we only discuss one special case: $R_{56}^{(1)} = R_{56}^{(3)}, h_1 = h_2 = h$. The treatment of more general cases with different signs and magnitudes of $R_{56}^{(1)}$ and $R_{56}^{(3)}$ and h_1 and h_2 is similar, but the same-signed $R_{56}^{(1)}$ and $R_{56}^{(3)}$ might be easier for a real lattice to fulfill. For example if $R_{56}^{(1)}, R_{56}^{(3)} > 0$, a possible realization of them are chicanes.

For the special case of $R_{56}^{(1)} = R_{56}^{(3)}, h_1 = h_2 = h$ and denote $\zeta_1 \equiv 1 + R_{56}^{(1)}h$, $\zeta_2 \equiv 2 + R_{56}^{(2)}h$, we then have

$$\mathbf{M}_R = \begin{pmatrix} \zeta_1\zeta_2 - 1 & \frac{\zeta_1^2\zeta_2 - 2\zeta_1}{h} \\ h\zeta_2 & \zeta_1\zeta_2 - 1 \end{pmatrix}. \tag{2.83}$$

The linear stability requires $|\zeta_1\zeta_2 - 1| < 1$, and the synchrotron tune is

$$\nu_s = \begin{cases} \frac{1}{2\pi} \arccos[\zeta_1\zeta_2 - 1] & \text{if } \frac{\zeta_1^2\zeta_2 - 2\zeta_1}{h} > 0, \\ 1 - \frac{1}{2\pi} \arccos[\zeta_1\zeta_2 - 1] & \text{if } \frac{\zeta_1^2\zeta_2 - 2\zeta_1}{h} < 0. \end{cases} \tag{2.84}$$

Here we give one example parameter set with a stable linear motion as shown in Table 2.2. According to the longitudinal Courant-Snyder functions given in Table 2.2 (note the values of β_z and the signs of α_z), the evolution of electron distribution in the longitudinal phase space around the ring (note the bunch lengths and orientations) is qualitatively shown in Fig. 2.6. If we can realize $\epsilon_z \lesssim 5$ pm in such a strong focusing ring, then we have $\sigma_z(s_{\text{rad}}) = \sqrt{\epsilon_z\beta_z(s_{\text{rad}})} \lesssim 3$ nm. We remind the readers that the contribution of modulators to longitudinal emittance should be carefully counted in a longitudinal strong focusing SSMB storage ring. Note that the energy chirp strength needed here is one order of magnitude smaller than the example of using a single RF or laser modulator to realize 3 nm as discussed just now. This benefit originates from a much compressed β_z at the radiator in a strong focusing ring. However, we recognize

Table 2.2 An example parameters set corresponding to the setup shown in Fig. 2.6. The subscripts $-/+$ means right in front and after the corresponding element

Parameter	Value
$R_{56}^{(1)}$	$15\,\mu m$
$R_{56}^{(2)}$	$-100\,\mu m$
$R_{56}^{(3)}$	$15\,\mu m$
h	$-5 \times 10^4\,m^{-1}$
ζ_1	0.25
ζ_2	7
ν_s	0.115
$\beta_z(s_{rad})$	$1.9\,\mu m$
$\beta_z(s_{RF})$	$121\,\mu m$
$\alpha_z(s_{rad})$	0
$\alpha_z(s_{RF1-})$	-7.9
$\alpha_z(s_{RF1+})$	-1.9
$\alpha_z(s_{RF2-})$	1.9
$\alpha_z(s_{RF2+})$	7.9

that the laser power needed (20 MW level if 270 nm-wavelength laser is applied) is still demanding, and here our primary goal is to present the principle based on which the interested readers can choose and optimize the parameters for their target applications. We will discuss in Chap. 3 the application of transverse-longitudinal coupling scheme to lower the requirement on the modulation laser power further, to make the optical cavity can be operated in CW mode, thus to improve the filling factor of electron beam in the ring and the average output radiation power.

2.1.6.2 Discussions

Here we make several observations from the above analysis and numerical example, which we believe are important. First, β_z in a longitudinal strong focusing ring can be at the same level of or even smaller than the ring $|R_{56} = -\eta C_0|$, while in a longitudinal weak focusing ring $\beta_z \gg |-\eta C_0|$. Therefore, the bunch length can thus be much smaller than that in a longitudinal weak focusing ring. This is the reason behind the application of longitudinal strong focusing in SSMB to realize extreme short bunches [28, 31]. We remind the readers that the longitudinal emittance of electron beam in a longitudinal strong focusing ring still cannot be smaller than that given in Eq. (2.63), due to the intrinsic partial phase slippage, thus the evolution of longitudinal beta function, in a dipole.

Second, β_z changes significantly around the ring in the longitudinal strong focusing regime. Therefore, the bunch length and beam orientation in the longitudinal phase space varies greatly around the ring, as shown qualitatively in Fig. 2.6. This means the adiabatic approximation cannot be applied for the longitudinal dimension anymore. Actually the adiabatic approximation also breaks down in the case corresponds to Fig. 2.3, where the change of β_z around the ring is significant although the total synchrotron phase advance per turn is small. Therefore, the global synchrotron tune is not a general criterion in the classification of whether the adiabatic approximation fails. The evolution of β_z is more relevant. The argument is based on the fact that R_{56} can be either positive or negative, therefore the local synchrotron phase advance can be either positive or negative. While in the transverse dimension, the drift length and betatron phase advance are always positive.

The breakdown of adiabatic approximation may have crucial impacts on the study of both the single-particle and collective effects. For linear single-particle dynamics, the longitudinal and transverse dimensions should be treated the same way on equal footing and SLIM formalism can be invoked. The treatment of nonlinear single-particle dynamics is more subtle as the longitudinal dynamics now is strongly chaotic. For the collective effects, many classical treatments should be re-evaluated and some new formalism needs to be developed. For example, the Haissinski equation [32] for calculating the equilibrium beam distortion cannot be applied directly then. Also, to our knowledge, there is no discussion on coherent synchrotron radiation (CSR)-induced microwave instability in a longitudinal strong focusing ring. The scaling law obtained in the longitudinal weak focusing [33] cannot be applied directly. 3D CSR effects and also the impact of bunch lengthening from transverse emittance on CSR needs more in-depth study. This is especially true for an SSMB ring, considering the fact that the beam width there is much larger than the microbunch length, while the contrary is true in a conventional ring. The contribution from horizontal emittance can easily dominate the bunch length at many places in an SSMB ring. This on the other hand, will be helpful to suppress unwanted CSR and may also be helpful in mitigating intrabeam scattering (IBS) [34, 35], as extreme short bunches occur only at limited locations. The IBS in a longitudinal strong focusing ring, and a general coupled lattice, also deserves special attention. To our knowledge, the IBS formalism of presented in Refs. [36, 37] can be applied for such purposes, as they are based on 6×6 general transport matrices. An IBS formalism can also be developed based on SLIM formalism [9], in which eigen analysis has been invoked and applies to 3D general coupled lattice with longitudinal strong focusing.

2.1.7 Thick-Lens Maps of a Laser Modulator

In the previous discussions, we have approximated the function of a laser modulator by a thin-lens RF-like kick. This means that we have ignored the phase slippage or R_{56} of the laser modulator itself. We need to know if this approximation is valid or under what circumstance we can use this approximation.

Here we first derive the phase slippage factor of the undulator and then get the thick-lens transfer matrix of a laser modulator. The path length of an electron with a relative energy deviation of δ wiggling in a planar undulator is

$$L(\delta) = \int_0^{L_u} \sqrt{1 + (x')^2} dz \approx \int_0^{L_u} \left[1 + \frac{1}{2} \left(\frac{K}{\gamma} \cos(k_u z) \right)^2 \right] dz$$

$$= \left[1 + \frac{1}{4} \frac{K^2}{\gamma_r^2 (1 + \delta)^2} \right] L_u \approx \left(1 - \frac{1}{2} \frac{K^2}{\gamma_r^2} \delta \right) L_u, \tag{2.85}$$

in which $k_u = 2\pi/\lambda_u$ is the undulator wavenumber, γ_r is the Lorentz factor corresponding to the resonant energy. The R_{56} of an undulator is then

$$R_{56} = \frac{L_u - L(\delta)}{\delta} + \frac{L_u}{\gamma_r^2} = \frac{L_u(1 + K^2/2)}{\gamma_r^2} = 2N_u \lambda_0, \tag{2.86}$$

where N_u is the number of undulator periods, $\lambda_0 = \frac{1+K^2/2}{2\gamma_r^2}\lambda_u$ is the central wavelength of the on-axis fundamental spontaneous radiation. As can be seen from Eq. (2.86), the undulator R_{56} is twice the slippage length of the undulator radiation.

As mentioned, the RF or laser modulator kick in linear approximation is like a longitudinal quadrupole and the R_{56} of the laser modulator is like the longitudinal drift space length of this longitudinal quadrupole. Assuming that the energy modulation is uniform along the undulator, then similar to the thick-lens quadrupole in the transverse dimension, we have the thick-lens transfer matrix of a laser modulator

$$\mathbf{M} = \begin{cases} \begin{pmatrix} \cos\left(\sqrt{-R_{56}h}\right) & \frac{R_{56}\sin\left(\sqrt{-R_{56}h}\right)}{\sqrt{-R_{56}h}} \\ \frac{h\sin\left(\sqrt{-R_{56}h}\right)}{\sqrt{-R_{56}h}} & \cos\left(\sqrt{-R_{56}h}\right) \end{pmatrix}, & \text{if } R_{56}h < 0, \\ \begin{pmatrix} \cosh\left(\sqrt{R_{56}h}\right) & \frac{R_{56}\sinh\left(\sqrt{R_{56}h}\right)}{\sqrt{R_{56}h}} \\ \frac{h\sinh\left(\sqrt{R_{56}h}\right)}{\sqrt{R_{56}h}} & \cosh\left(\sqrt{R_{56}h}\right) \end{pmatrix}, & \text{if } R_{56}h > 0. \end{cases} \tag{2.87}$$

A laser modulator in linear approximation is therefore like a thick-lens quadrupole in the longitudinal dimension. A thin-lens approximation is applicable when $|R_{56}h| \ll 1$.

After discussing the linear map, now we take into account the fact that the modulation waveform of a laser is actually sinusoidal. In principle, we can get an approximate analytical nonlinear thick-lens transfer map of the laser modulator using the techniques of drift and kick and Lie algebra [38, 39], by slicing the interaction into several smaller pieces and concatenate the maps of thin-lens kickes and drift spaces. Here we use a more straightforward method, i.e., to implement a symplectic kick map as below in a numerical code, to give the readers a picture. The kick map implemented in the code is as follows

$$\text{for } i = 1:1:N_u$$
$$z = z + \lambda_0 \delta$$
$$\delta = \delta + A_i \sin(k_L z) \tag{2.88}$$
$$z = z + \lambda_0 \delta$$
$$\text{end}$$

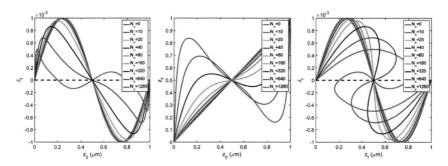

Fig. 2.7 Impact of modulator undulator period number N_u (thus R_{56}) on the single-pass modulation process in a laser modulator. The injected beam is "$\delta = 0$ and $z \in [0, \lambda_0]$". (x, y) in the figures from left to right: ($z_{\text{entrance}}, \delta_{\text{exit}}$); ($z_{\text{entrance}}, z_{\text{exit}}$); ($z_{\text{exit}}, \delta_{\text{exit}}$). Parameters used: $\lambda_L = 1\ \mu m$, $A = 1 \times 10^{-3}$

In other words, we have split the undulator into N_u small "$\frac{1}{2}$dispersion + modulation + $\frac{1}{2}$dispersion". For a plane wave $A_i = \frac{A}{N_u}$ with A being the total energy modulation strength, while for a Gaussian laser beam A_i is a function of i.

To simplify the analysis, in the example numerical simulation, we consider the case of a plane wave, i.e., $A_i = \frac{A}{N_u}$. First, we want to see how the single-pass modulation waveform is like based on Eq. (2.88). We choose parameters $\lambda_L = \lambda_0 = 1\ \mu m$, $A = 1 \times 10^{-3}$. The single-pass modulation of a line beam, with "$\delta = 0$ and $z \in [0, \lambda_0]$", as a function of N_u, namely R_{56}, of the undulator is shown in Fig. 2.7. When changing N_u, we keep A unchanged. As can be seen, the waveform deviates from *sine* wave when N_u increases. The beam distribution in phase space at the undulator exit (right sub-figure of Fig. 2.7) is similar to the beam de-coherence in an RF bucket.

Now we consider the multi-pass cases, i.e., we consider the impact of modulator R_{56} on the phase space bucket. But here we do simulation only for the longitudinal weak focusing with a single laser modulator, as we only aim to give the readers a picture about such impact. We use parameters of $\lambda_L = \lambda_0 = 1\ \mu m$, $A = 1 \times 10^{-3}$, $C_0 = 100\ m$, $\eta = 5 \times 10^{-7}$ in the simulation, and choose to observe the beam opposite the modulator center where $\alpha_z = 0$ with $N_u = 0, 40, 320$, respectively. Note that η is the phase slippage factor of the whole ring, including the modulator. When changing the undulator priod number N_u, we keep η a constant. The results are shown in Fig. 2.8. It can be seen that the modulator R_{56} only distorts the bucket slightly when N_u is 40. But when N_u is as large as 320, it will have a profound effect on the longitudinal phase space topology. Its impact on longitudinal strong focusing is more subtle as the particle motion in a longitudinal strong focusing ring is strongly chaotic if the nonlinear modulation waveform is taken into account. The study of such effect can refer more straightforwardly to numerical simulations. Besides, the undulator R_{56} could also have an impact on the coherent radiation induced collective instability in the laser modulator [40, 41].

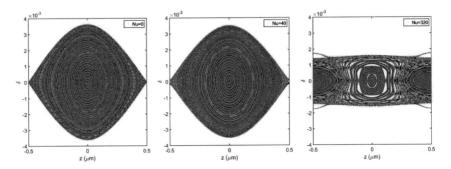

Fig. 2.8 Impact of modulator undulator period number N_u (thus R_{56}) on the longitudinal phase space bucket in longitudinal weak focusing regime. Parameters used: $\lambda_L = 1 \ \mu m$, $A = 1 \times 10^{-3}$, $C_0 = 100$ m, $\eta = 5 \times 10^{-7}$

2.2 Nonlinear Longitudinal Dynamics

After resolving the issue of quantum diffusion of z by means of dedicated lattice design, we can then apply the quasi-isochronous or low-phase slippage method to realize short bunches in SSMB. However, the phase slippage is actually a function of the particle energy

$$\eta(\delta) = \eta_0 + \eta_1\delta + \eta_2\delta^2 + ... \tag{2.89}$$

When η_0 is sufficiently small, the higher-order terms in Eq. (2.89) may become relevant or even dominant, and the beam dynamics can be significantly different from those in a linear-phase slippage state. Proper application of dedicated sextupoles and octupoles may be needed to control these higher-order terms.

The beam dynamics of the quasi-isochronous rings have been studied by many authors [15, 16, 42]. Here, we wish to emphasize two points that have not been well investigated before and might be important, for example, in the SSMB proof-of-principle experiment to be introduced in Chap. 5 and the longitudinal dynamic aperture optimization in SSMB.

2.2.1 For High-Harmonic Bunching

For seeding techniques such as coherent harmonic generation (CHG) [43] and high-gain harmonic generation (HGHG) [44, 45], it seems that to date, linear phase slippage or R_{56} has been applied for microbunching formation. Here, we wish to point out that one can actually take advantage of the nonlinearity of the phase slippage for high harmonic generation. Intuitively, this is because a sinusoidal energy modulation followed by a nonlinear phase slippage can lead to a distorted current distribution, which, in some cases, can lead to large bunching at a specific harmonic number.

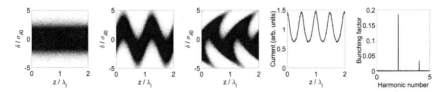

Fig. 2.9 The beam evolution in longitudinal phase space, final current distribution and bunching factor, when $\eta(\delta) = \eta_1\delta$ is used for microbunching in CHG or HGHG, as modeled by Eq. (2.90)

Figure 2.9 shows an example simulation of using $\eta(\delta) = \eta_1\delta$ for microbunching. It can be seen that there is significant bunching in the second and fourth harmonics, while no bunching is produced in the fundamental and third harmonics. The reason can be found from the following derivation of the bunching factor.

The microbunching process in the case of a single energy modulation followed by a dispersion section, as that in CHG and HGHG, can be modeled as

$$\delta' = \delta + A\sin(k_L z),$$
$$z' = z - \eta(\delta')C_0\delta', \qquad (2.90)$$

where $k_L = 2\pi/\lambda_L$ is the wavenumber of the modulation laser, A is the electron energy modulation strength induced by the laser. The bunching factor at the wavenumber of k is defined as

$$b(k) = \int_{-\infty}^{\infty} dz\, e^{-ikz}\rho(z), \qquad (2.91)$$

where $\rho(z)$ is the normalized longitudinal density distribution of the electron beam satisfying $\int_{-\infty}^{\infty} dz\rho(z) = 1$. According to Liouville's theorem, we have $dzd\delta = dz'd\delta'$. Therefore, the bunching factor can be calculated in accordance with the initial distribution of the particles $\rho_0(z, \delta)$ as

$$b(k) = \int_{-\infty}^{\infty}\int_{-\infty}^{\infty} dzd\delta\, \rho_0(z, \delta)e^{-ikz'(z,\delta)}. \qquad (2.92)$$

Here we consider the simple case of $\eta(\delta) \equiv \eta_0 + \eta_1\delta$, then

$$b(k) = \int_{-\infty}^{\infty}\int_{-\infty}^{\infty} dzd\delta\, \rho_0(z, \delta)e^{ik\left(\eta_0 C_0\delta + \eta_1 C_0\delta^2 + \frac{\eta_1 C_0 A^2}{2}\right)}$$
$$e^{ik\left[-z + (\eta_0 C_0 A + 2\eta_1 C_0\delta A)\sin(k_L z) - \frac{\eta_1 C_0 A^2}{2}\cos(2k_L z)\right]}. \qquad (2.93)$$

Adopting the notation $Y \equiv k(\eta_0 C_0 A + 2\eta_1 C_0\delta A)$, $Z \equiv -k\eta_1 C_0 A^2/2$, and using the mathematical identity $e^{ix\sin(k_L z)} = \sum_{n=-\infty}^{\infty} e^{ink_L z}J_n[x]$, we have

$$e^{-ikz+iY\sin(k_L z)+iZ\cos(2k_L z)} = \sum_{p=-\infty}^{\infty} \sum_{q=-\infty}^{\infty} J_p[Y]J_q[Z]\exp\left(-i\left[(k-(p-2q)k_L)z - q\frac{\pi}{2}\right]\right).$$

$$(2.94)$$

If the initial beam is much longer than the laser wavelength, and considering that $\frac{1}{2\pi}\int_{-\infty}^{\infty} e^{-i\omega t}d\omega = \delta(t)$, where $\delta(t)$ is the Dirac delta function, the bunching factor will not vanish only if $k = (p-2q)k_L$. The bunching factor at the n-th harmonic of the modulation laser is then

$$b_n = \int_{-\infty}^{\infty} d\delta\, e^{ink_L\left(\eta_0 C_0\delta + \eta_1 C_0\delta^2 + \frac{\eta_1 C_0 A^2}{2}\right)} \rho_0(\delta) \sum_{m=-\infty}^{\infty} J_{n+2m}[Y]J_m[Z]. \qquad (2.95)$$

Here we consider the simple case of an initial Gaussian energy distribution $\rho_0(\delta) = \frac{1}{\sqrt{2\pi}\sigma_\delta}\exp\left(-\frac{\delta^2}{2\sigma_\delta^2}\right)$, where σ_δ is the initial RMS energy spread. If $\eta_1 = 0$, then $Y = nk_L\eta_0 C_0 A$, $Z = 0$, and $\sum_{m=-\infty}^{\infty} J_{n+2m}[Y]J_m[Z] = J_n[nk_L\eta_0 C_0 A]$, and we have

$$b_n = J_n[nk_L\eta_0 C_0 A]\exp\left[-\frac{(nk_L\eta_0 C_0\sigma_\delta)^2}{2}\right], \qquad (2.96)$$

which is a familiar result for HGHG [44] if we adopt the notation $R_{56} = -\eta_0 C_0$. If $\eta_0 = 0$, then $Y = 2nk_L\eta_1 C_0\delta A$, $Z = -nk_L\eta_1 C_0 A^2/2$, meaning that we have

$$b_n = \frac{1}{\sqrt{2\pi}\sigma_\delta}\int_{-\infty}^{\infty} d\delta\, \exp\left[ink_L\left(\eta_1 C_0\delta^2 + \frac{\eta_1 C_0 A^2}{2}\right)\right]\exp\left(-\frac{\delta^2}{2\sigma_\delta^2}\right)\sum_{m=-\infty}^{\infty} J_{n+2m}[Y]J_m[Z].$$

$$(2.97)$$

The two exponential terms in the integral are even functions of δ, while $J_{n+2m}[Y]J_m[Z]$ is an odd function of δ when n is odd; thus, b_n is nonzero only for an even n. This is why bunching occurs only in the second and fourth harmonics but not in the fundamental and third harmonics when we use $\eta(\delta) = \eta_1\delta$ for microbunching, as shown in Fig. 2.9.

Following the derivations and according to the relation

$$\cos^n(x) = \frac{1}{2^{n-1}}\sum_{m=(n+1)/2}^{n}\binom{n}{m}\cos(2m-n)x, \qquad (2.98)$$

it can be seen that the energy modulation at the fundamental frequency can be cast into $[i \times (n-2p) + j \times (n-2q)]$-th harmonic bunching through the term $\eta_{n-1}\delta^n$ in the function of $\eta(\delta)$. For an odd n, bunching at all harmonic numbers are possible, while for an even n, only bunching at the even harmonic numbers is possible. The optimal bunching condition for a specific harmonic requires the matching of $\eta(\delta)$ with the energy modulation strength. However, the analytical formula for the bunching factor will become increasingly involved with more higher-order terms of the phase slippage considered. Thus, it would be better to refer to numerical code to calculate and optimize the bunching factor directly for a specific application case. For storage rings, another relevant point is that the distribution of the particle energy in the nonlinear phase slippage state may also have an impact on the high harmonic generation, and this phenomenon is also easier to be studied by means of numerical simulation.

The approach of applying a nonlinear phase slippage for high harmonic bunching can be considered to share some similarity with echo-enabled harmonic generation (EEHG) [46, 47], in which the sinusoidal energy modulation and dispersion in the first stage can be viewed as the source of the distorted current distribution in the second stage of modulation and dispersion for microbunching. We have also noticed the work on optimizing the nonlinearity of the dispersion to increase the bunching factor for EEHG [48]. Based on similar considerations, tricks can also be applied on the energy-modulation waveform using different harmonics of the modulation laser, for example, forming a sawtooth waveform to boost bunching, as will be discussed in Chap. 3.

2.2.2 For Longitudinal Dynamic Aperture

Similar to the transverse dimension, there is a region in the longitudinal phase space outside of which particle motion is not bounded and can be lost in a ring. We refer this stable region as the longitudinal dynamic aperture. Here in this section, we want to show that, by properly tailoring the nonlinear phase slippage, the longitudinal dynamic aperture can be enlarged significantly compared to the case of a pure linear phase slippage. Only symplectic dynamics is considered in this discussion.

2.2.2.1 Longitudinal Weak Focusing

The longitudinal dynamics of a particle in a ring with a single RF can be modeled by the kick map

$$\begin{cases} \delta_{n+1} = \delta_n + A[\sin(k_{RF}z_n) - \sin\phi_s], \\ z_{n+1} = z_n - \eta(\delta_{n+1})C_0\delta_{n+1}, \end{cases} \tag{2.99}$$

where $A\sin\phi_s = U_0/E_0$ where U_0 is the radiation loss of a particle per turn. For the case of longitudinal weak focusing, the kicik map can be approximated by differentiation and Hamiltonian formalism can be invoked for the analysis. Denote $\phi \equiv k_{RF}z$, then the equation of motion is

$$\begin{cases} \frac{d\phi}{dt} = -\frac{k_{RF}\eta(\delta_{n+1})C_0}{T_0}\delta = \frac{\partial\mathcal{H}}{\partial\delta}, \\ \frac{d\delta}{dt} = \frac{A}{T_0}(\sin\phi - \sin\phi_s) = -\frac{\partial\mathcal{H}}{\partial\phi}, \end{cases} \tag{2.100}$$

with T_0 being the revolution period of the particle in the ring. For $\eta(\delta) = \eta_0 + \eta_1\delta + \eta_2\delta^2$, the corresponding Hamiltonian is

$$\mathcal{H}(\phi, \delta) = -\omega_{RF}\left(\frac{1}{2}\eta_0\delta^2 + \frac{1}{3}\eta_1\delta^3 + \frac{1}{4}\eta_2\delta^4\right) + \frac{A}{T_0}\left[\cos\phi - \cos\phi_s + (\phi - \phi_s)\sin\phi_s\right].$$
$$(2.101)$$

In writing down the closed-form Hamiltonian, we have implicitly assumed that the motion is integrable, i.e., there is no chaos. But we need to keep in mind that the dynamics dictated by Eq. (2.99) is actually chaotic even with a linear phase slippage [13]. But here we ignore this subtle point as the chaotic layer is very thin in the longitudinal weak focusing regime. We remind the readers that the chaotic dynamics, for example the bucket bifurcation, can actually also be applied for ultrashort bunch generation [49].

To analyze the motion, we need to find the fixed points of the system

$$\begin{cases} \frac{\partial H}{\partial \phi} = 0 \\ \frac{\partial H}{\partial \delta} = 0 \end{cases} \implies \begin{cases} \sin\phi_s - \sin\phi = 0, \\ \delta\eta(\delta) = 0. \end{cases} \qquad (2.102)$$

To determine whether a fixed point is stable or not, we need to check the trace of the Jacobian matrix around the fixed point. If $\eta(\delta) = \eta_0$, there are two sets of fixed points:

$$\begin{cases} \text{SFP}: \ (\phi_s, 0), \\ \text{UFP}: \ (\pi - \phi_s, 0), \end{cases} \qquad (2.103)$$

in which SFP stands for stable fixed point while UFP stands for unstable fixed point. If $\eta(\delta) = \eta_0 + \eta_1\delta$, there are four sets of fixed points:

$$\begin{cases} \text{SFP}: \ (\phi_s, 0), \ \left(\pi - \phi_s, -\frac{\eta_0}{\eta_1}\right), \\ \text{UFP}: \ (\pi - \phi_s, 0), \ \left(\phi_s, -\frac{\eta_0}{\eta_1}\right). \end{cases} \qquad (2.104)$$

If $\eta(\delta) = \eta_0 + \eta_1\delta + \eta_2\delta^2$, there are six sets of fixed points

$$\begin{cases} \text{SFP}: \ (\phi_s, 0), \ \left(\pi - \phi_s, -\frac{\eta_1}{2\eta_2} + \sqrt{\left(\frac{\eta_1}{2\eta_2}\right)^2 - \frac{\eta_0}{\eta_2}}\right), \ \left(\pi - \phi_s, -\frac{\eta_1}{2\eta_2} - \sqrt{\left(\frac{\eta_1}{2\eta_2}\right)^2 - \frac{\eta_0}{\eta_2}}\right), \\ \text{UFP}: \ (\pi - \phi_s, 0), \ \left(\phi_s, -\frac{\eta_1}{2\eta_2} + \sqrt{\left(\frac{\eta_1}{2\eta_2}\right)^2 - \frac{\eta_0}{\eta_2}}\right), \ \left(\phi_s, -\frac{\eta_1}{2\eta_2} - \sqrt{\left(\frac{\eta_1}{2\eta_2}\right)^2 - \frac{\eta_0}{\eta_2}}\right). \end{cases}$$
$$(2.105)$$

To see the impact of η_1 and η_2 on the longitudinal phase space bucket, some numerical simulations are conducted. We choose the observation point at the middle of the RF, where $\alpha_z = 0$ and the beam distribution in the longitudinal phase space is upright. The results of the impact of η_1 and η_2 on longitudinal dynamical aperture are shown in Figs. 2.10 and 2.11, respectively. Note that in the plots, we have used the longitudinal coordinate z rather than the phase ϕ.

As we can see in Fig. 2.10, the emergence of η_1 will make the bucket asymmetric in δ, which is as expected as the $\eta = \eta_0 + \eta_1\delta$ is asymmetric in δ. In both directions

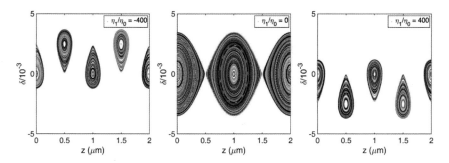

Fig. 2.10 The impact of η_1 on the longitudinal phase space bucket in the longitudinal weak focusing regime. Simulation parameters: $\lambda_{RF} = 1\ \mu m$, $A = 1 \times 10^{-3}$, $C_0 = 100$ m, $\eta_0 = 5 \times 10^{-7}$

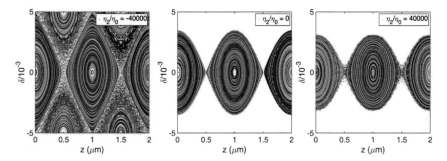

Fig. 2.11 The impact of η_2 on the longitudinal phase space bucket in the longitudinal weak focusing regime. Simulation parameters: $\lambda_{RF} = 1\ \mu m$, $A = 1 \times 10^{-3}$, $C_0 = 100$ m, $\eta_0 = 5 \times 10^{-7}$

(positive or negative), the bucket size shrinks with the increase of η_1 and the bucket becomes like an upright α-shape, so they are usually referred to as α-buckets. Note that we can also classify the bucket to be an RF-bucket or an α-bucket according to whether $\delta = 0$ or $\eta(\delta) = 0$ at the bucket center, respectively. Such a classification is more reasonable from beam dynamics consideration. α-bucket is also a method to generate short bunch and there are many interesting beam dynamics issues related to such buckets [16].

However, the impact of η_2 is different. As can be seen from the simulation results presented in Fig. 2.11, the bucket is still symmetric in δ as expected. Besides, when $\frac{\eta_2}{\eta_0} < 0$, the stable region of the bucket can be even larger than the case without η_2. We will see later that in the case of longitudinal strong focusing, such observation can be even more notable. Therefore, we can tailor the phase slippage factor as a function of energy in the case of longitudinal strong focusing to enlarge the longitudinal dynamic aperture. This is very helpful as usually the longitudinal dynamic aperture in a longitudinal strong focusing ring is not a trivial issue and needs to be optimized to guarantee a sufficient quantum lifetime for example.

2.2.2.2   Longitudinal Strong Focusing

For a longitudinal strong focusing ring, the particle motion is strongly chaotic and not integrable, and we cannot get a closed-form Halmitonian for analysis anymore. Therefore, we use numerical simulations to study the dynamics directly. Instead of a comprehensive investigations, here in this section we aim to give some qualitative remarks on the role of the nonlinear phase slippage, i.e., a proper tailoring of the nonlinear phase slippage can enlarge the longitudinal dynamic aperture significantly.

Let us use the schematic layout shown in Fig. 2.6 and parameters choices given in Table 2.2 as an example for illustration. Note that the modulation wavelength used here is $\lambda_{RF} = 1$ μm. We choose to observe the beam at the radiator center, and suppose the ring is symmetric with respect to the radiator. The one-turn kick map is then

$$
\begin{aligned}
z &= z + R_{56}^{(1)}\delta, \\
\delta &= \delta + h/k_{RF}\sin(k_{RF}z), \\
z &= z + R_{56}^{(2)}\delta, \\
\delta &= \delta + h/k_{RF}\sin(k_{RF}z), \\
z &= z + R_{56}^{(1)}\delta.
\end{aligned}
\tag{2.106}
$$

As we aim to present the main physical picture, here we only consider the nonlinearity of the main ring first, i.e., $R_{56}^{(2)}(\delta) = -C_0\left(\eta_0 + \eta_1\delta + \eta_2\delta^2\right)$. $R_{56}^{(1)}$ and $R_{56}^{(3)}$ in principle can also be a function of δ. The simulation results are shown in Figs. 2.12 and 2.13.

From Fig. 2.12, we know that, like that in the weak focusing case, η_1 makes the bucket asymmetric in δ and shrinks the bucket size whether η_1 is positive or negative. From Fig. 2.13, we can see that when $\frac{\eta_2}{\eta_0} < 0$, a proper η_2 can help to merge the island buckets with the main bucket and broaden the stable region of the phase space, i.e.,

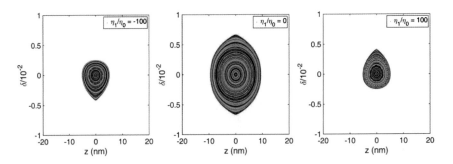

Fig. 2.12 The impact of η_1 on the longitudinal phase space bucket in the longitudinal strong focusing regime. Simulation parameters: $\lambda_{RF} = 1$ μm, $h = -50000$ m^{-1}, $R_{56}^{(1)} = 15$ μm, $C_0 = 100$ m, $\eta_0 = 1 \times 10^{-6}$

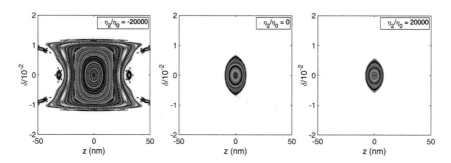

Fig. 2.13 The impact of η_2 on the longitudinal phase space bucket in the longitudinal strong focusing regime. Simulation parameters: $\lambda_{RF} = 1 \ \mu m$, $h = -50000 \ m^{-1}$, $R_{56}^{(1)} = 15 \ \mu m$, $C_0 = 100 \ m$, $\eta_0 = 1 \times 10^{-6}$

the longitudinal dynamic aperture, significantly. A proper η_2 makes the amplitude dependent tune shift favorable for the motion to be stable. Note that the fixed points of the island buckets may not have period-1 but period-n stability.

References

1. Deng XJ, Chao AW, Feikes J, Huang WH, Ries M, Tang CX (2020) Single-particle dynamics of microbunching. Phys Rev Accel Beams 23:044002
2. Deng XJ, Chao AW, Huang WH, Tang CX (2021) Courant-snyder formalism of longitudinal dynamics. Phys Rev Accel Beams 24:094001
3. Deng XJ, Chao AW, Feikes J, Hoehl A, Huang WH, Klein R, Kruschinski A, Li J, Ries M, Tang CX (2023) Breakdown of classical bunch length and energy spread formula in a quasi-isochronous electron storage ring. Phys Rev Accel Beams 26:054001
4. Zhang Y, Deng XJ, Pan ZL, Li ZZ, Zhou KS, Huang WH, Li RK, Tang CX, Chao AW (2021) Ultralow longitudinal emittance storage rings. Phys Rev Accel Beams 24:090701
5. Sands M (1970) Physics of electron storage rings: an introduction. Technical report, Stanford Linear Accelerator Center, Calif
6. Shoji Y, Tanaka H, Takao M, Soutome K (1996) Longitudinal radiation excitation in an electron storage ring. Phys Rev E 54:R4556–R4559
7. Soutome K, Takao M, Tanaka H, Shoji Y (1998) Longitudinal radiation excitation of quasi-isochronus storage ring"new subaru". In: Proceedings of the 6th European particle accelerator conference (EPAC'98), Stockholm, Sweden, 1998. Institute of Physics Publishing, Bristol and Philadelphia, pp 1008–1010
8. Biscari C (2005) Bunch length modulation in highly dispersive storage rings. Phys Rev ST Accel Beams 8:091001
9. Chao AW (1979) Evaluation of beam distribution parameters in an electron storage ring. J Appl Phys 50(2):595–598
10. Courant ED, Snyder HS (1958) Theory of the alternating-gradient synchrotron. Ann Phys (NY) 3(1):1–48
11. Wang MC, Uhlenbeck GE (1945) On the theory of the brownian motion ii. Rev Mod Phys 17:323–342
12. Chandrasekhar S (1943) Stochastic problems in physics and astronomy. Rev Mod Phys 15:1–89

13. Chirikov BV (1979) A universal instability of many-dimensional oscillator systems. Phys Rep 52(5):263–379
14. Klein R, Brandt G, Fliegauf R, Hoehl A, Müller R, Thornagel R, Ulm G, Abo-Bakr M, Feikes J, Hartrott MV, Holldack K, Wüstefeld G (2008) Operation of the metrology light source as a primary radiation source standard. Phys Rev ST Accel Beams 11:110701
15. Feikes J, von Hartrott M, Ries M, Schmid P, Wüstefeld G, Hoehl A, Klein R, Müller R, Ulm G (2011) Metrology light source: the first electron storage ring optimized for generating coherent thz radiation. Phys Rev ST Accel Beams 14:030705
16. Ries M (2014) Nonlinear momentum compaction and coherent synchrotron radiation at the metrology light source. Dissertation, Humboldt University of Berlin. PhD thesis, Humboldt-Universität zu Berlin, Berlin
17. Safranek J (1997) Experimental determination of storage ring optics using orbit response measurements. Nucl Instrum Methods Phys Res A 388(1):27–36
18. Shoji Y (2004) Bunch lengthening by a betatron motion in quasi-isochronous storage rings. Phys Rev ST Accel Beams 7:090703
19. Klein R, Mayer T, Kuske P, Thornagel R, Ulm G (1997) Beam diagnostics at the bessy i electron storage ring with compton backscattered laser photons: measurement of the electron energy and related quantities. Nucl Instrum Methods Phys Res A 384(2–3):293–298
20. Campbell N (1909) The study of discontinuous phenomena. Proc Cambridge Phil Soc 15:117–136
21. Sands M (1955) Synchrotron oscillations induced by radiation fluctuations. Phys Rev 97:470–473
22. Rice SO (1944) Mathematical analysis of random noise. Bell Syst Tech J 23(3):282–332
23. Jowett JM (1987) Introductory statistical mechanics for electron storage rings. In: AIP conference proceedings, vol 153. American Institute of Physics, pp 864–970
24. Pan Z (2020) Research on optimization and design of advanced laser-driving storage ring. PhD thesis, Tsinghua University, Beijing
25. Pan Z, Rui T, Wan W, Chao A, Deng X, Zhang Y, Huang W, Tang C (2019) A storage ring design for steady-state microbunching to generate coherent euv light source. In: Proceedings of the 39th international free electron laser conference (FEL'19), Hamburg, Germany, 2019. JACoW, Geneva, pp 700–703
26. Tang C, Deng X (2022) Steady-state micro-bunching accelerator light source. Acta Phys Sin 71:152901
27. Chao A (2022) Focused laser, unpublished note
28. Chao A, Granados E, Huang X, Ratner D, Luo H-W (2016) High power radiation sources using the steady-state microbunching mechanism. In: Proceedings of the 7th international particle accelerator conference (IPAC'16), Busan, Korea, 2016. JACoW, Geneva, pp 1048–1053
29. Christofilos N (1950) Focusing system for ions and electrons. US Patent, vol 2, pp 736–799
30. Courant ED, Livingston MS, Snyder HS (1952) The strong-focusing synchroton-a new high energy accelerator. Phys Rev 88:1190–1196
31. Zhang Y (2022) Research on longitudinal strong focusing SSMB ring. PhD thesis, Tsinghua University, Beijing
32. Haissinski J (1973) Exact longitudinal equilibrium distribution of stored electrons in the presence of self-fields. Il Nuovo Cimento B 18(1):1971–1996
33. Bane KLF, Cai Y, Stupakov G (2010) Threshold studies of the microwave instability in electron storage rings. Phys Rev ST Accel Beams 13:104402
34. Piwinski A (1974) Intra-beam-scattering. In: Proceedings of the 9th international conference on high energy accelerators, Stanford, p 405
35. Bjorken JD, Mtingwa SK (1982) Intrabeam scattering. Part Accel 13(FERMILAB-PUB-82-47-THY):115–143
36. Nash B (2006) Analytical approach to Eigen-emittance in storage rings. PhD thesis, Stanford University, Stanford
37. Kubo K, Oide K (2001) Intrabeam scattering in electron storage rings. Phys Rev ST Accel Beams 4:124401

38. Dragt AJ (2021) Lie methods for nonlinear dynamics with applications to accelerator physics
39. Chao AW (2022) Special topics in accelerator physics. World Scientific
40. Tsai C-Y, Chao AW, Jiao Y, Luo H-W, Ying M, Zhou Q (2021) Coherent-radiation-induced longitudinal single-pass beam breakup instability of a steady-state microbunch train in an undulator. Phys Rev Accel Beams 24:114401
41. Tsai C-Y (2022) Theoretical formulation of multiturn collective dynamics in a laser cavity modulator with comparison to robinson and high-gain free-electron laser instability. Phys Rev Accel Beams 25:064401
42. Pellegrini C, Robin D (1991) Quasi-isochronous storage ring. Nucl Instrum Methods Phys Res, Sect A 301(1):27–36
43. Girard B, Lapierre Y, Ortega JM, Bazin C, Billardon M, Elleaume P, Bergher M, Velghe M, Petroff Y (1984) Optical frequency multiplication by an optical klystron. Phys Rev Lett 53:2405–2408
44. Yu LH (1991) Generation of intense uv radiation by subharmonically seeded single-pass free-electron lasers. Phys Rev A 44:5178–5193
45. Yu L-H, Babzien M, Ben-Zvi I, DiMauro L, Doyuran A, Graves W, Johnson E, Krinsky S, Malone R, Pogorelsky I et al (2000) High-gain harmonic-generation free-electron laser. Science 289(5481):932–934
46. Stupakov G (2009) Using the beam-echo effect for generation of short-wavelength radiation. Phys Rev Lett 102:074801
47. Xiang D, Stupakov G (2009) Echo-enabled harmonic generation free electron laser. Phys Rev ST Accel Beams 12:030702
48. Stupakov G, Zolotorev M (2011) Using laser harmonics to increase bunching factor in eehg. In: Proceedings of the 33th international free electron laser conference (FEL'11), Shanghai, China, 2011, pp 45–48
49. Jiao Y, Ratner DF, Chao AW (2011) Terahertz coherent radiation from steady-state microbunching in storage rings with x-band radio-frequency system. Phys Rev ST Accel Beams 14:110702

Chapter 3
SSMB Transverse-Longitudinal Coupling Dynamics

After dedicated efforts devoted to the longitudinal dynamics to realize an ultrasmall longitudinal emittance and ultrashort bunch length for coherent radiation generation, we need to make sure that the coupling arising from transverse dynamics does not degrade or even destroy the longitudinal fine structures. Such an argument is based on the observation that the transverse beam size in an SSMB ring can be orders of magnitude larger than the desired microbunch length. This is the basic motivation for us to investigate the transverse-longitudinal coupling (TLC) dynamics. In this chapter, we start from the linear TLC and then investigate the nonlinear TLC dynamics. For the linear dynamics, first we analyze the passive bunch lengthening induced by bending magnets. We then emphasize the fact that TLC can actually be actively applied for efficient bunch compression and high harmonic generation when the transverse emittance is small. We present three theorems on the application of such TLC schemes, with their implications discussed. Further, we have analyzed the contribution of modulators to the vertical emittance from quantum excitation, to obtain a self-consistent evaluation of the required modulation laser power when applying these coupling schemes in a storage ring. The theorems and related analysis provide the theoretical basis for the application of TLC in SSMB to lower the requirement on the modulation laser power, by taking advantage of the fact that the vertical emittance in a planar ring is rather small. Based on the investigations, we have presented example parameters sets for the envisioned SSMB storage ring to generate high-power EUV and soft X-ray radiation at the end of this dissertation. The relation between our TLC analysis and the transverse-longitudinal emittance exchange is also briefly discussed. For the nonlinear dynamics, we present the analysis and the first experiment proof of a second-order TLC effect on the equilibrium beam parameters, which can help to improve the stable beam current and coherent radiation power of a ring working in quasi-isochronous regime. Parts of the work presented in this chapter have been published in Refs. [1–4].

© The Author(s) 2024

X. Deng, *Theoretical and Experimental Studies on Steady-State Microbunching*,
Springer Theses, https://doi.org/10.1007/978-981-99-5800-9_3

3.1 Linear Transverse-Longitudinal Coupling Dynamics

3.1.1 Passive Bunch Lengthening

In a linear transport line without bending magnets, the transverse and longitudinal motions are decoupled in a first-order approximation. However, the situation changes when there are bending magnets. Particles with different horizontal (vertical) positions and angles will pass through the horizontal (vertical) bending magnets along different paths, resulting in differences in the longitudinal coordinate. The transverse motion can thus be coupled to the longitudinal dimension. When traversing the bending magnets, particles with different energies will also pass along different paths and exit with different horizontal (vertical) positions and angles. The longitudinal motion can thus also be coupled to the transverse dimension. The physical pictures of the linear TLC introduced by the bending magnets are shown in Fig. 3.1. Although this passive TLC is a well-understood effect [5–9], here we present a concise analysis of this effect with an emphasis on its vital role in microbunching formation and transportation for both the transient and steady-state cases.

We start with a planar x-y uncoupled lattice and assume that the RF cavities are placed at dispersion-free locations. We temporarily ignoring the vertical dimension, and use the state vector $\mathbf{X} = (x, x', z, \delta)^T$. The subscripts $5, 6$ are used for z, δ for consistency with literature. Hereafter, the subscript x in this section is omitted unless necessary. As introduced in Sect. 2.1.1, the betatron coordinate, defined by $\mathbf{X}_\beta = \mathbf{B}\mathbf{X}$, is first used to parametrize the transport matrix in a diagonal form. The transport matrix of $\mathbf{X}_\beta$ from s_1 to s_2 is then

$$\mathbf{M}_\beta(s_2, s_1) = \begin{pmatrix} \mathbf{M}_{x_\beta}(s_2, s_1) & \mathbf{0} \\ \mathbf{0} & \mathbf{M}_{z_\beta}(s_2, s_1) \end{pmatrix}. \tag{3.1}$$

Following Courant and Snyder [10], we write $\mathbf{M}_{x_\beta}(s_2, s_1)$ as

$$\mathbf{M}_{x_\beta}(s_2, s_1) = \mathbf{A}^{-1}(s_2)\mathbf{T}(s_2, s_1)\mathbf{A}(s_1), \tag{3.2}$$

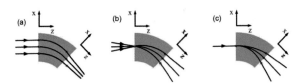

Fig. 3.1 Linear transverse-longitudinal coupling induced by a bending magnet. Particles with different horizontal positions (**a**) and angles (**b**) pass the horizontal bending magnet along different paths, resulting in longitudinal coordinate differences. Particles with different energies (**c**) also pass the horizontal bending magnet along different paths, resulting in horizontal position and angle differences

with

$$A(s_i) = \begin{pmatrix} \frac{1}{\sqrt{\beta(s_i)}} & 0 \\ \frac{\alpha(s_i)}{\sqrt{\beta(s_i)}} & \sqrt{\beta(s_i)} \end{pmatrix} \tag{3.3}$$

and

$$T(s_2, s_1) = \begin{pmatrix} \cos\psi_{12} & \sin\psi_{12} \\ -\sin\psi_{12} & \cos\psi_{12} \end{pmatrix}, \tag{3.4}$$

where

$$\psi_{12} = \psi_2 - \psi_1 = \int_{s_1}^{s_2} \frac{1}{\beta(s)} ds \tag{3.5}$$

is the betatron phase advance from s_1 to s_2. For $M_{z_\beta}(s_2, s_1)$, the expression is similar to $M_{x_\beta}(s_2, s_1)$, but note that if we want to calculate the synchrotron phase advance similar to Eq. (3.5), the distance s should be replaced by the effective longitudinal drift space, i.e., $F = -\tilde{\eta}(s_2, s_1)C_0$ defined in Eq. (2.50). If there is no RF cavity between s_1 and s_2, we have

$$M_{z_\beta}(s_2, s_1) = \begin{pmatrix} 1 & F(s_2, s_1) \\ 0 & 1 \end{pmatrix}. \tag{3.6}$$

The transport matrix of X from s_1 to s_2 is then

$$M(s_2, s_1) = B^{-1}(s_2)M_\beta(s_2, s_1)B(s_1). \tag{3.7}$$

After some straightforward algebra, $M(s_2, s_1)$ can be expressed as

$$M(s_2, s_1) = \begin{pmatrix} R_{11} & R_{12} & 0 & D_2 - R_{11}D_1 - R_{12}D_1' \\ R_{21} & R_{22} & 0 & D_2' - R_{21}D_1 - R_{22}D_1' \\ R_{51} & R_{52} & 1 & F - R_{51}D_1 - R_{52}D_1' \\ 0 & 0 & 0 & 1 \end{pmatrix},$$

$$R_{11} = \sqrt{\frac{\beta_2}{\beta_1}}[\cos\psi_{12} + \alpha_1 \sin\psi_{12}],$$

$$R_{12} = \sqrt{\beta_1\beta_2} \sin\psi_{12},$$

$$R_{21} = -\frac{1}{\sqrt{\beta_1\beta_2}}[(1 + \alpha_1\alpha_2)\sin\psi_{12} - (\alpha_1 - \alpha_2)\cos\psi_{12}], \tag{3.8}$$

$$R_{22} = \sqrt{\frac{\beta_1}{\beta_2}}[\cos\psi_{12} - \alpha_2 \sin\psi_{12}],$$

$$R_{51} = R_{21}D_2 - R_{11}D_2' + D_1',$$

$$R_{52} = -R_{12}D_2' + R_{22}D_2 - D_1.$$

This matrix can then be used to analyze both the transient and steady-state cases of TLC. Note that for a given lattice, F is a function of the initial dispersion and dispersion angle (D_1, D_1') at s_1, although the transfer map $\mathbf{M}(s_2, s_1)$ for the state vector $\mathbf{X}$ is not, as the transport matrix is fixed once the lattice is given. This dependence is a result of the matrix parametrization.

We consider first the influence of the betatron oscillation on the longitudinal coordinate. With the help of the Courant-Snyder parametrization, the betatron oscillation position and angle at the starting point s_1 can be expressed as

$$x_1 = \sqrt{2J\beta_1} \cos \psi_1,$$
$$x_1' = -\sqrt{2J/\beta_1}(\alpha_1 \cos \psi_1 + \sin \psi_1), \tag{3.9}$$

where $J = \frac{1}{2} \left(\gamma x^2 + 2\alpha x x' + \beta x'^2 \right)$ is the betatron invariant or action of the particle. The longitudinal coordinate displacement relative to the ideal particle due to the betatron oscillation from s_1 to s_2 is then

$$\Delta z = R_{51} x_1 + R_{52} x_1' = \sqrt{2J\mathcal{H}_1} \sin(\psi_1 - \chi_1) - \sqrt{2J\mathcal{H}_2} \sin(\psi_2 - \chi_2), \tag{3.10}$$

where to obtain the final concise result, D and D' have been expressed in terms of the chromatic $\mathcal{H}$-function and the chromatic phase χ, defined as

$$D = \sqrt{\mathcal{H}\beta} \cos \chi,$$
$$D' = -\sqrt{\mathcal{H}/\beta} \left(\alpha \cos \chi + \sin \chi \right), \tag{3.11}$$

where $\mathcal{H} = \gamma D^2 + 2\alpha D D' + \beta D'^2$. If there is no dipole kick between point 1 and point 2, $\mathcal{H}$ stays constant and $\chi_2 - \chi_1 = \psi_2 - \psi_1$, which means $\Delta z = 0$. Physically, this means the contribution of transverse emittance to the bunch length does not change during drifting or experiencing quadrupole kicks, as these manipulations only affect the beam distribution in the transverse phase space. This argument can also be clearly observed in Fig. 2.3c.

From Eq. (3.10), the root-mean-square (RMS) value of the transient bunch lengthening of a longitudinal slice from s_1 to s_2 caused by this linear TLC can be calculated to be

$$\sigma_{\Delta z} = \sqrt{\epsilon_x \left[\mathcal{H}_1 + \mathcal{H}_2 - 2\sqrt{\mathcal{H}_1 \mathcal{H}_2} \cos \left(\Delta \psi_{21} - \Delta \chi_{21} \right) \right]}. \tag{3.12}$$

The RMS bunch lengthening of an electron beam longitudinal slice after n complete revolutions in a ring, due to betatron oscillation, is then

$$\sigma_{\Delta z} = 2\sqrt{\epsilon_x \mathcal{H}_x} \left| \sin(n\pi \nu_x) \right|. \tag{3.13}$$

The above equations can be used to explain the dependence of the coherent synchrotron radiation (CSR) repetition rate on the betatron tune in the bunch slicing experiment reported in Ref. [8]. A similar approach can be applied to analyze microbunching preservation with beam deflection, for example in FEL multiplexing. These equations are also useful for evaluating the influence of the coupling effect in the SSMB proof-of-principle experiment [11, 12], which is to be presented in Chap. 5.

If the particle starts with a relative energy deviation of δ, then

$$\Delta z = R_{51}x_1 + R_{52}x_1' + (F - R_{51}D_1 - R_{52}D_1')\delta$$
$$= \sqrt{2J\mathcal{H}_1}\sin(\psi_1 - \chi_1) - \sqrt{2J\mathcal{H}_2}\sin(\psi_2 - \chi_2) + F\delta. \tag{3.14}$$

Note that in Eq. (3.14), the betatron invariant and phase should be calculated according to

$$x_\beta = x - D\delta = \sqrt{2J\beta}\cos\psi,$$
$$x_\beta' = x' - D'\delta = -\sqrt{2J/\beta}(\alpha\cos\psi + \sin\psi),$$
$$J = \frac{1}{2}\left(\gamma x_\beta^2 + 2\alpha x_\beta x_\beta' + \beta x_\beta'^2\right). \tag{3.15}$$

For a periodic system, if we observe the particle at the same place n periods later, then

$$\Delta z = \sqrt{2J\mathcal{H}}\left[\sin(\psi - \chi) - \sin(\psi + 2n\pi\nu_x - \chi)\right] - n\eta C_0\delta, \tag{3.16}$$

where $2\pi\nu_x$ is the horizontal betatron phase advance per period.

We can also obtain the equilibrium second moments in a storage ring by following the Courant-Snyder parametrization one step further. The result is the same with Eq. (2.18) obtained by SLIM. As can be seen from Eqs. (2.18) and (2.19), if there is only passive TLC introduced by bending magnet, the transverse emittance can lengthen the bunch at places where $\mathcal{H}_x \neq 0$,

$$\sigma_z = \sqrt{\epsilon_z\beta_z + \epsilon_x\mathcal{H}_x}. \tag{3.17}$$

Similarly the energy spread can broaden the beam width at places where $D \neq 0$,

$$\sigma_x = \sqrt{\epsilon_x\beta_x + \epsilon_z\gamma_z D^2} = \sqrt{\epsilon_x\beta_x + \sigma_\delta^2 D^2}. \tag{3.18}$$

To give the readers a more concrete feeling about the bunch lengthening from this passive TLC, we have presented in Fig. 2.3c some calculations based on the MLS lattice. As can be seen, indeed that the coupling from horizontal emittance can contribute significantly, or even dominant the bunch length at places where $\mathcal{H}_x$ is large. This observation will especially be true in an SSMB ring, where the transverse size is much larger than the microbunch length. Therefore, the dispersion and dispersion angle should be controlled in precision at places where ultrashort bunch is desired, for example at the radiator. The bunch lengthening from the transverse

Fig. 3.2 Beam current distributions at places with different $\mathcal{H}_x$. Bunch length in an SSMB ring can easily be dominated by the transverse emittance in places where $\mathcal{H}_x \neq 0$

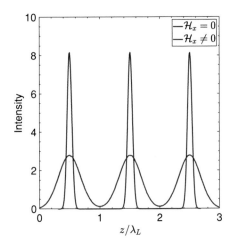

emittance will make the current distribution in an SSMB storage ring less sharp and more like a coasting beam as places where $\mathcal{H}_x \neq 0$, as shown in Fig. 3.2. Here we make a remark that this coupling effect may be helpful for suppressing unwanted CSR and may mitigate the intrabeam scattering (IBS) in SSMB or other applications, as extremely short bunches emerge only at dispersion-free locations.

3.1.2   Coupling for Harmonic Generation and Bunch Compression

The analysis in the above section may lead us to conclude that TLC always lengthens the bunch and degrades the microbunching. This, however, is not true, as the above analysis is based on the assumption of a planar x-y uncoupled lattice with only the passive coupling induced by the bending magnets. In addition to this passive coupling, an RF cavity (laser modulator in SSMB) placed at a dispersive location, a transverse deflecting RF cavity, etc., are other sources of coupling that can be used for subtle manipulation of particle beam in 6D phase space. In fact, one can take advantage of TLC for efficient harmonic generation or bunch compression when the transverse emittance is small. The reason is that there is some flexibility in tailoring the projection of the three eigen emittances of a beam into different physical dimensions, although their values cannot be changed in a linear symplectic lattice. Here in this and the following sections we investigate the active application of TLC for harmonic generation and bunch compression by taking advantage of the fact that the vertical emittance of an electron beam in a planar x-y uncoupled ring is rather small.

Fig. 3.3 A schematic layout of applying y-z coupling for bunch compression

3.1.2.1 Problem Definition

Let us first define the problem we are trying to solve. We use y-z coupling as an example for the analysis, since we aim to exploit the small vertical emittance. The case of x-z coupling is similar. Suppose the beam at the entrance of the bunch compression section is y-z decoupled, i.e., its second moments matrix is

$$\Sigma_i = \begin{pmatrix} \epsilon_y\beta_{yi} & -\epsilon_y\alpha_{yi} & 0 & 0 \\ -\epsilon_y\alpha_{yi} & \epsilon_y\gamma_{yi} & 0 & 0 \\ 0 & 0 & \epsilon_z\beta_{zi} & -\epsilon_z\alpha_{zi} \\ 0 & 0 & -\epsilon_z\alpha_{zi} & \epsilon_z\gamma_{zi} \end{pmatrix}, \tag{3.19}$$

where α, β and γ are the Courant-Snyder functions [10], the subscript i means initial, and ϵ_y and ϵ_z are the eigen emittances of the beam corresponding to the vertical and longitudinal mode, respectively. For the application of TLC for bunch compression, it means that the final bunch length at the radiator depends only on the vertical emittance ϵ_y and not on the longitudinal one ϵ_z. The magnet lattices are all planar and x-y decoupled.

The schematic layout of a TLC-based bunch compression section is shown in Fig. 3.3. We divide such a section into three parts, with their transfer matrices given by

$$\mathbf{M}_1 = \begin{pmatrix} r_{33} & r_{34} & 0 & d \\ r_{43} & r_{44} & 0 & d' \\ r_{53} & r_{54} & 1 & r_{56} \\ 0 & 0 & 0 & 1 \end{pmatrix}, \ \mathbf{M}_2 = \text{Modulation kick map}, \ \mathbf{M}_3 = \begin{pmatrix} R_{33} & R_{34} & 0 & D \\ R_{43} & R_{44} & 0 & D' \\ R_{53} & R_{54} & 1 & R_{56} \\ 0 & 0 & 0 & 1 \end{pmatrix},$$

$$r_{53} = r_{43}d - r_{33}d', \ r_{54} = -r_{34}d' + r_{44}d, \ r_{33}r_{44} - r_{34}r_{43} = 1,$$
$$R_{53} = R_{43}D - R_{33}D', \ R_{54} = -R_{34}D' + R_{44}D, \ R_{33}R_{44} - R_{34}R_{43} = 1,$$
$$\tag{3.20}$$

with $\mathbf{M}_1$ representing "from entrance to modulator", $\mathbf{M}_2$ representing "modulation kick" and $\mathbf{M}_3$ representing "modulator to radiator". Note that $\mathbf{M}_1$ and $\mathbf{M}_3$ are in their general thick-lens form as analyzed in last section. The transfer matrix from the entrance to the radiator is then

$$\mathbf{T} = \mathbf{M}_3\mathbf{M}_2\mathbf{M}_1. \tag{3.21}$$

From the problem definition, for $\sigma_z(\text{Rad})$ to be independent of ϵ_z, we need

$$T_{55} = 0,$$
$$T_{56} = 0. \tag{3.22}$$

3.1.2.2　Three Theorems on Transverse-Longitudinal Coupling

Given the above problem definition, we have three theorems which dictate the relation between the modulator kick strength with the optical functions at the modulator and radiator, respectively.

Theorem one: If

$$\mathbf{M}_2 = \begin{pmatrix} 1 & 0 & 0 & 0 \\ 0 & 1 & 0 & 0 \\ 0 & 0 & 1 & 0 \\ 0 & 0 & h & 1 \end{pmatrix}, \tag{3.23}$$

which corresponds to the case of a normal RF or a TEM00 mode laser modulator, then

$$h^2(\text{Mod})\mathcal{H}_y(\text{Mod})\mathcal{H}_y(\text{Rad}) \geq 1. \tag{3.24}$$

Theorem two: If

$$\mathbf{M}_2 = \begin{pmatrix} 1 & 0 & 0 & 0 \\ 0 & 1 & t & 0 \\ 0 & 0 & 1 & 0 \\ t & 0 & 0 & 1 \end{pmatrix}, \tag{3.25}$$

which corresponds to the case of a transverse deflecting RF or a TEM01 mode laser modulator or other schemes for angular modulation, then

$$t^2(\text{Mod})\beta_y(\text{Mod})\mathcal{H}_y(\text{Rad}) \geq 1. \tag{3.26}$$

Theorem three: If

$$\mathbf{M}_2 = \begin{pmatrix} 1 & 0 & k & 0 \\ 0 & 1 & 0 & 0 \\ 0 & 0 & 1 & 0 \\ 0 & -k & 0 & 1 \end{pmatrix}, \tag{3.27}$$

whose physical correspondence is not as straightforward as the previous two cases, then

$$k^2(\text{Mod})\gamma_y(\text{Mod})\mathcal{H}_y(\text{Rad}) \geq 1. \tag{3.28}$$

3.1.2.3　Proof

Here we present the details for the proof of Theorem one. The proof of the other two is just similar. From the problem definition, for $\sigma_z(\text{Rad})$ to be independent of ϵ_z, we need

$$T_{55} = hR_{56} + 1 = 0,$$
$$T_{56} = dR_{53} + d'R_{54} + r_{56}(hR_{56} + 1) + R_{56} = 0. \tag{3.29}$$

Note that the harmonic generation schemes in FEL like phase-merging enhanced harmonic generation (PEHG) [13, 14] and angular dispersion-induced microbunching (ADM) [15] can be viewed as specific examples of the above general relations [2]. Under the conditions of Eq. (3.29), we have

$$\mathbf{T} = \begin{pmatrix} \mathbf{A} & \mathbf{B} \\ \mathbf{C} & \mathbf{E} \end{pmatrix}, \tag{3.30}$$

with

$$\mathbf{A} = \begin{pmatrix} r_{33}R_{33} + r_{43}R_{34} + r_{53}hD & r_{34}R_{33} + r_{44}R_{34} + r_{54}hD \\ r_{33}R_{43} + r_{43}R_{44} + r_{53}hD' & r_{34}R_{43} + r_{44}R_{44} + r_{54}hD' \end{pmatrix},$$
$$\mathbf{B} = \begin{pmatrix} hD & dR_{33} + d'R_{34} + (r_{56}h + 1)D \\ hD' & dR_{43} + d'R_{44} + (r_{56}h + 1)D' \end{pmatrix},$$
$$\mathbf{C} = \begin{pmatrix} r_{33}R_{53} + r_{43}R_{54} & r_{34}R_{53} + r_{44}R_{54} \\ r_{53}h & r_{54}h \end{pmatrix}, \tag{3.31}$$
$$\mathbf{E} = \begin{pmatrix} 0 & 0 \\ h & r_{56}h + 1 \end{pmatrix}.$$

The bunch length squared at the modulator and the radiator are

$$\sigma_z^2(\text{Mod}) = \epsilon_z \left(\beta_{zi} - 2\alpha_{zi}r_{56} + \gamma_{zi}r_{56}^2 \right) + \epsilon_y \frac{\left(\beta_{yi}r_{53} - \alpha_{yi}r_{54} \right)^2 + r_{54}^2}{\beta_{yi}}$$
$$= \epsilon_z\beta_z(\text{Mod}) + \epsilon_y\mathcal{H}_y(\text{Mod}), \tag{3.32}$$
$$\sigma_z^2(\text{Rad}) = \epsilon_y \frac{\left(\beta_{yi}T_{53} - \alpha_{yi}T_{54} \right)^2 + T_{54}^2}{\beta_{yi}} = \epsilon_y\mathcal{H}_y(\text{Rad}).$$

According to Cauchy-Schwarz inequality, we have

$$h^2(\text{Mod})\mathcal{H}_y(\text{Mod})\mathcal{H}_y(\text{Rad}) = h^2 \frac{\left[(\beta_{yi}r_{53} - \alpha_{yi}r_{54})^2 + r_{54}^2 \right] \left[(\beta_{yi}T_{53} - \alpha_{yi}T_{54})^2 + T_{54}^2 \right]}{\beta_{yi}} \frac{}{\beta_{yi}}$$
$$\geq \frac{h^2}{\beta_y^2} \left[-(\beta_{yi}r_{53} - \alpha_{yi}r_{54})T_{54} + r_{54}(\beta_{yi}T_{53} - \alpha_{yi}T_{54}) \right]^2$$
$$= (T_{53}r_{54}h - T_{54}r_{53}h)^2 = (T_{53}T_{64} - T_{54}T_{63})^2. \tag{3.33}$$

The equality holds when $\frac{-(\beta_{yi}r_{53} - \alpha_{yi}r_{54})}{T_{54}} = \frac{r_{54}}{(\beta_{yi}T_{53} - \alpha_{yi}T_{54})}$. The symplecticity of $\mathbf{T}$ requires that $\mathbf{TST}^T = \mathbf{S}$, where $\mathbf{S} = \begin{pmatrix} \mathbf{J} & 0 \\ 0 & \mathbf{J} \end{pmatrix}$ and $\mathbf{J} = \begin{pmatrix} 0 & 1 \\ -1 & 0 \end{pmatrix}$, so we have

$$\begin{pmatrix} \mathbf{AJA}^T + \mathbf{BJB}^T & \mathbf{AJC}^T + \mathbf{BJE}^T \\ \mathbf{CJA}^T + \mathbf{EJB}^T & \mathbf{CJC}^T + \mathbf{EJE}^T \end{pmatrix} = \begin{pmatrix} \mathbf{J} & \mathbf{0} \\ \mathbf{0} & \mathbf{J} \end{pmatrix}. \tag{3.34}$$

As shown in Eq. (3.31), $\mathbf{E} = \begin{pmatrix} 0 & 0 \\ h & r_{56}h + 1 \end{pmatrix}$, then $\mathbf{EJE}^T = \begin{pmatrix} 0 & 0 \\ 0 & 0 \end{pmatrix}$. Therefore,

$$\mathbf{CJC}^T = \mathbf{J}, \tag{3.35}$$

which means $\mathbf{C}$ is also a symplectic matrix. So we have

$$T_{53}T_{64} - T_{54}T_{63} = \det(\mathbf{C}) = 1, \tag{3.36}$$

where $\det(\mathbf{C})$ means the determinant of $\mathbf{C}$. The theorem is thus proven.

3.1.2.4 Dragt's Minimum Emittance Theorem

Theorem one in Eq. (3.24) can also be expressed as

$$|h(\text{Mod})| \geq \frac{\epsilon_y}{\sqrt{\epsilon_y \mathcal{H}_y(\text{Mod})}\sqrt{\epsilon_y \mathcal{H}_y(\text{Rad})}} = \frac{\epsilon_y}{\sigma_{zy}(\text{Mod})\sigma_z(\text{Rad})}. \tag{3.37}$$

Note that in the above formula, $\sigma_{zy}(\text{Mod})$ means the bunch length at the modulator contributed from the vertical emittance ϵ_y. So given a fixed ϵ_y and desired $\sigma_z(\text{Rad})$, a smaller $h(\text{Mod})$, i.e., a smaller RF gradient or modulation laser power ($P_{\text{laser}} \propto |h(\text{Mod})|^2$), means a larger $\mathcal{H}_y(\text{Mod})$, thus a longer $\sigma_{zy}(\text{Mod})$, is needed. As $|h(\text{Mod})|\sigma_z(\text{Mod})$ quantifies the energy spread introduced by the modulation kick, we thus also have

$$\sigma_z(\text{Rad})\sigma_\delta(\text{Rad}) \geq \epsilon_y. \tag{3.38}$$

Similarly for Theorem two and three, we have

$$|t(\text{Mod})| \geq \frac{\epsilon_y}{\sigma_{y\beta}(\text{Mod})\sigma_z(\text{Rad})}, \tag{3.39}$$

and

$$|k(\text{Mod})| \geq \frac{\epsilon_y}{\sigma_{y'\beta}(\text{Mod})\sigma_z(\text{Rad})}, \tag{3.40}$$

respectively, and also Eq. (3.38). Note that in the above formulas, the vertical beam size or divergence at the modulator contains only the betatron part, i.e., that from the vertical emittance ϵ_y.

Equation (3.38) is actually a manifestation of the classical uncertainty principle [16], which states that

$$\Sigma_{11}\Sigma_{22} \geq \epsilon_{\min}^2,$$
$$\Sigma_{33}\Sigma_{44} \geq \epsilon_{\min}^2, \qquad (3.41)$$
$$\Sigma_{55}\Sigma_{66} \geq \epsilon_{\min}^2,$$

in which $\epsilon_{\min}$ is the minimum one among the three eigen emittances $\epsilon_{I,II,III}$. In our bunch compression case, we assume that ϵ_y is the smaller one compared to ϵ_z. Actually there is a stronger inequality compared to the classical uncertainty principle, i.e., the minimum emittance theorem [16], which states that the projected emittance cannot be smaller than the minimum one among the three eigen emittances,

$$\epsilon_{x,pro}^2 = \Sigma_{11}\Sigma_{22} - \Sigma_{12}^2 \geq \epsilon_{\min}^2,$$
$$\epsilon_{y,pro}^2 = \Sigma_{33}\Sigma_{44} - \Sigma_{34}^2 \geq \epsilon_{\min}^2, \qquad (3.42)$$
$$\epsilon_{z,pro}^2 = \Sigma_{55}\Sigma_{66} - \Sigma_{56}^2 \geq \epsilon_{\min}^2.$$

3.1.3 Normal RF or TEM00 Mode Laser for Coupling

Now we investigate in more detail about the application of TLC in SSMB for bunch compression, using a TEM00 mode laser modulator for the modulation kick. This belongs to the category of Theorem one.

3.1.3.1 Physical Picture

According to Theorem one, given a vertical emittance ϵ_y and modulation kick strength h, in principle we can realize as short σ_z(Rad) as we want by lengthening σ_{zy}(Mod). In other words, we can lower $\mathcal{H}_y$(Rad) by increasing $\mathcal{H}_y$(Mod). However, such great flexibility of a TLC coupling scheme is not obtained without sacrifice. For a premicrobunched beam, and considering that the modulation waveform is actually a nonlinear sinusoidal, a bunch lengthening at the modulator will result in bunching factor degradation at the radiator. Another key point is that the modulator itself will contribute to the vertical emittance through quantum excitation since it is placed at a place where $\mathcal{H}_y \neq 0$. We will elaborate these points more in this section.

To give the readers a better picture before going into the mathematical details, here we summarize in Fig. 3.4 the main information to be presented in this section: (i) Compared to bunch compression or harmonic generation scheme in longitudinal dimension alone like high-gain harmonic generation (HGHG), TLC schemes like PEHG or ADM can reduce the required energy chirp strength, to realize the same bunch length compression ratio or harmonic generation number, when the transverse emittance is small. (ii) This lowering of energy chirp strength is realized through the bunch lengthening from transverse emittance at the modulator, which can degrade the bunching factor at the radiator for a pre-microbunched beam due to the nonlinear

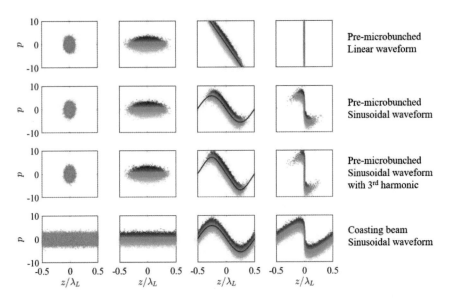

Fig. 3.4 Application of TLC for bunch compression and harmonic generation, using a TEM00 mode laser modulator. Parameters used in this example plot: $\lambda_L = 1064$ nm, $\sigma_{zi} = 30$ nm and $\sigma_{zf} = 3$ nm for the case of a pre-microbunched beam, $\sigma_{\delta i} = 3 \times 10^{-4}, \sigma_{yi} = 2\,\mu m, \sigma_{y'i} = 1\,\mu rad$. The figures show the beam distribution evolution in the longitudinal phase space. Depending on the specific lattice scheme, the different colors in the plot correspond to different particle vertical positions, angles, or combination of them. The modulation waveforms are shown in the figure as the red curves

nature of the *sine* modulation. (iii) Addition of the RF or laser harmonics is an effective way to mitigate this bunching factor degradation by broadening the linear zone of the modulation waveform.

3.1.3.2 Bunching Factor

Now we derive the bunching factor degradation at the radiator due to the bunch lengthening at the modulator, using ADM as an example. PEHG has a similar result as we have proven in the last section the general theorem of bunch lengthening in this kind of TLC schemes. Thin-lens kick maps in the last section are again used for the analysis, but now with the fact that the modulation waveform is a nonlinear *sine* taken into account.

Putting in the optimized bunch compression conditions for ADM, namely $hR_{56} + 1 = 0$ and $-d'D + R_{56} = 0$, and using the mathematical identity $e^{ia\sin(b)} = \sum_{m=-\infty}^{\infty} e^{imb} J_m[a]$, the final bunching factor at the n-th laser harmonic in ADM is

$$b_n = \sum_{m=-\infty}^{\infty} J_m(n) \int_{-\infty}^{\infty}\int_{-\infty}^{\infty}\int_{-\infty}^{\infty} dy_i dy_i' dz_i e^{-ink_L\left[-\frac{y_i'}{hd'}+\left(1-\frac{m}{n}\right)(d'y_i+z_i)\right]} f_i(y_i, y_i', z_i).$$

(3.43)

For a coasting beam, $\left\langle e^{-ink_L\left[\left(1-\frac{m}{n}\right)(d'y_i+z_i)\right]}\right\rangle$ will be non-zero only if $m = n$, where the bracket $\langle \cdots \rangle$ means the average over all the particles. Therefore,

$$b_{n,\text{coasting}} = J_n(n)\exp\left[-\left(nk_L\sigma_z(\text{Rad})\right)^2/2\right].$$

(3.44)

Note that $\sigma_z(\text{Rad}) = |D|\sigma_{y'i} = \sqrt{\epsilon_y \mathcal{H}_y(\text{Rad})}$ in this section follows the definition in the linear matrix analysis of the previous section, and does not represent the real bunch length at the radiator considering the nonlinear modulation waveform. Note also that considering the nonlinear *sine* modulation waveform, the optimal microbunching condition for a specific harmonic is slightly different from our simplified linear analysis, and $J_n(n)$ in Eq. (3.44) should be replaced by $J_n(-nR_{56}h)$. In the following discussions, we will use the simplified optimal bunch compression conditions, as the main physics is the same.

For a pre-microbunched beam, $\left\langle e^{-ink_L\left[\left(1-\frac{m}{n}\right)(d'y_i+z_i)\right]}\right\rangle$ will be non-zero for all m, thus

$$b_{n,\text{pre-microbunch}} = \left(\sum_{m=-\infty}^{\infty} J_m(n)\exp\left[-\left((n-m)k_L\sigma_z(\text{Mod})\right)^2/2\right]\right)$$
$$\exp\left[-\left(nk_L\sigma_z(\text{Rad})\right)^2/2\right],$$

(3.45)

with $\sigma_z(\text{Mod}) = \langle d'y_i + z_i \rangle = \sqrt{\epsilon_y \mathcal{H}_y(\text{Mod}) + \epsilon_z \beta_z(\text{Mod})}$. Note that the bunch length $\sigma_z(\text{Mod})$ here contains contribution from both ϵ_y and ϵ_z.

Now we first investigate two limiting cases. If $\sigma_z(\text{Mod}) = 0$, then we have

$$b_n = \left(\sum_{m=-\infty}^{\infty} J_m(n)\right)\exp\left[-\left(nk_L\sigma_z(\text{Rad})\right)^2/2\right] = \exp\left[-\left(nk_L\sigma_z(\text{Rad})\right)^2/2\right].$$

(3.46)

This result is the same as that assuming the modulation waveform is linear. Second, if $\sigma_z(\text{Mod})$ is much longer than the modulation laser wavelength, i.e., $k_L\sigma_z(\text{Mod}) \gg 1$, then the summation terms in Eq. (3.45) will be nonzero only for $m = n$ and we arrive at the same result as the coasting beam case Eq. (3.44) as expected.

Now we conduct a bit more general discussion. Compared to the linear modulation case, the reduction factor of the bunching factor Eq. (3.45) is

$$R_n = \sum_{m=-\infty}^{\infty} J_m(n)\exp\left[-\left((n-m)k_L\sigma_z(\text{Mod})\right)^2/2\right].$$

(3.47)

Fig. 3.5 Flat contour plot for the bunching factor reduction factor $|R_n|$ of Eq. (3.47) as a function of the harmonic number n and the modulation wavelength-normalized bunch length at the modulator $k_L \sigma_z(\text{Mod})$

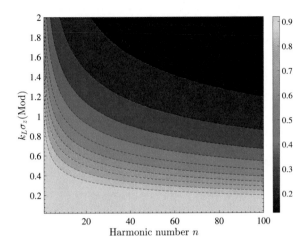

Figure 3.5 shows the flat contour plot for the bunching factor reduction factor $|R_n|$ of Eq. (3.47) as a function of the harmonic number n and the modulation wavelength-normalized bunch length at the modulator $k_L\sigma_z(\text{Mod})$. As can be seen from the figure, the bunch lengthening at the modulator indeed degrades the bunching factor at the radiator, due to the nonlinearity nature of *sine* modulation. The longer this bunch lengthening, the more degradation of the bunching factor. The higher the harmonic number, the more significant the impact is. The limit of R_n with an infinite long $\sigma_z(\text{Mod})$ is $J_n(n)$. Equation (3.47) and Fig. 3.5 is the general result of this bunching factor degradation analysis. We emphasize the fact that the discussion of bunching factor degradation is more relevant for a pre-microbunhed beam, like that in some SSMB scenarios, and is generally not an issue for a coasting beam where the bunch duration is much longer than the modulation wavelength, like that in an FEL.

As the decrease of bunching factor originates from the nonlinearity of the *sine* modulation, we expect that this reduction will be less if we make the modulation waveform more like linear, for example by adding a third-harmonic RF or laser to broaden the linear zone of the modulation waveform, as also suggested before in Refs. [17, 18]. The energy modulation then becomes $\delta = \delta + \frac{h_1}{k_L}\sin(k_L z) + \frac{h_3}{3k_L}\sin(3k_L z)$. The optimized bunch compression conditions for ADM are now $(h_1 + h_3)R_{56} + 1 = 0$ and $-d'D + R_{56} = 0$. The n-th laser harmonic bunching factor at the radiator is then

$$b_{n,\text{coasting}} = \sum_{m_1+3m_3=n} J_{m_1}\left(\frac{h_1}{h_1+h_3}n\right) J_{m_3}\left(\frac{h_3}{h_1+h_3}\frac{n}{3}\right) \exp\left[-\left(nk_L\sigma_z(\text{Rad})\right)^2/2\right]$$

(3.48)

for a coasting beam, and

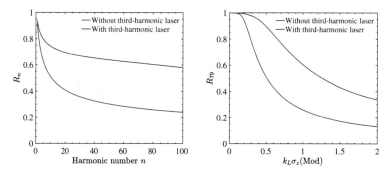

Fig. 3.6 Left: the bunching factor reduction factor $|R_n|$ of Eq. (3.50) as a function of the harmonic number n for $k_L\sigma_z(\text{Mod}) = 1$, with $h_3 = 0$ (red) and $h_3 = -0.15h_1$ (blue), respectively. Right: the bunching factor reduction factor $|R_n|$ of Eq. (3.50) as a function of the modulation wavelength-normalized bunch length at the modulator $k_L\sigma_z(\text{Mod})$ for $n = 79$, with $h_3 = 0$ (red) and $h_3 = -0.15h_1$ (blue), respectively. $n = 79$ corresponds to the case for example a modulation wavelength of $\lambda_L = 1064$ nm and a radiation wavelength of $\lambda_R = \lambda_L/79 = 13.5$ nm

$$
b_{n,\text{pre-microbunch}} = \sum_{m_1=-\infty}^{\infty} \sum_{m_3=-\infty}^{\infty} J_{m_1}\left(\frac{h_1}{h_1+h_3}n\right) J_{m_3}\left(\frac{h_3}{h_1+h_3}\frac{n}{3}\right)
$$
$$
\exp\left[-((n-m_1-3m_3)k_L\sigma_z(\text{Mod}))^2/2\right]\exp\left[-(nk_L\sigma_z(\text{Rad}))^2/2\right]
$$
$$(3.49)$$

for a pre-microbunched beam. Therefore, the reduction factor of the bunching factor Eq. (3.49), compared to the linear modulation case, is now

$$
R_n = \sum_{m_1=-\infty}^{\infty} \sum_{m_3=-\infty}^{\infty} J_{m_1}\left(\frac{h_1}{h_1+h_3}n\right) J_{m_3}\left(\frac{h_3}{h_1+h_3}\frac{n}{3}\right)
$$
$$
\exp\left[-((n-m_1-3m_3)k_L\sigma_z(\text{Mod}))^2/2\right].
$$
$$(3.50)$$

The limit of R_n with an infinite long $\sigma_z(\text{Mod})$ is $\sum_{m_1+3m_3=n} J_{m_1}\left(\frac{h_1}{h_1+h_3}n\right) J_{m_3}$ $\left(\frac{h_3}{h_1+h_3}\frac{n}{3}\right)$. It is straightforward to generalize the above derivation and result to the case of adding more laser harmonics.

Now we can use the above formula of R_n to do comparison between the cases with and without the third-harmonic laser. If $h_3 = 0$, then Eq. (3.50) reduces to Eq. (3.47). As can be seen in Fig. 3.6, indeed addition of a third-harmonic laser is effective in mitigating the bunching factor degradation arising from the bunch lengthening at the modulator.

3.1.3.3   Contribution of Modulators to Vertical Emittance and Scaling of Required Modulation Laser Power

We have stated that the main motivation of applying TLC scheme for bunch compression in SSMB is to lower the requirement on the modulation laser power P_L. This is based on the fact that the vertical emittance ϵ_y in a planar x-y uncoupled ring is rather small. However, since the modulator in this TLC scheme is placed at a dispersive location, i.e., $\mathcal{H}_y(\text{Mod}) \neq 0$, therefore quantum excitation at the modulator will also contribute to ϵ_y. With this consideration taken into account, below we try to give a self-consistent analysis of the required modulation laser power P_L in these TLC schemes.

To make sure that the TLC-based bunch compression can repeat turn-by-tun in a ring, usually two laser modulators are placed upstream and downstream of the radiator, respectively, to form a pair. The lattice scheme between these two modulators can either be a symmetric one, or a reversible seeding one. In both cases, the chromatic function $\mathcal{H}_y$ at two modulators are identical. Assuming the modulator undulator is planar, the contribution of these two modulators to ϵ_y is then

$$
\begin{aligned}
\Delta\epsilon_y(\text{Mod, QE}) &= 2 \times \frac{55}{96\sqrt{3}} \frac{\alpha_F \lambda_e^2 \gamma^5}{\alpha_V} \int_0^{L_u} \frac{\mathcal{H}_y(\text{Mod})}{|\rho(s)|^3} ds \\
&= 2 \times \frac{55}{96\sqrt{3}} \frac{\alpha_F \lambda_e^2 \gamma^5}{\alpha_V} \frac{\mathcal{H}_y(\text{Mod})}{\rho_{0\text{Mod}}^3} \frac{4}{3\pi} L_u,
\end{aligned}
\tag{3.51}
$$

with the vertical damping constant

$$
\alpha_V \approx \frac{1}{2}\frac{U_0}{E_0} \approx \frac{1}{2}C_\gamma \frac{E_0^3}{\rho_{\text{ring}}} = \frac{1}{2}C_\gamma \times 0.2998 \, B_{\text{ring}}[\text{T}]E_0^2[\text{GeV}]
\tag{3.52}
$$

where $C_\gamma = 8.85 \times 10^{-5} \frac{\text{m}}{\text{GeV}^3}$, ρ_{ring} is the bending radius of dipoles in the ring, and $\rho_{0\text{Mod}}$ is the minimum being radius corresponding to the peak magnetic flux density $B_{0\text{Mod}}$ of the modulator. Note that in the above analysis we have ignored the contribution of the dispersive lattice sections upstream and downstream of the modulators to the vertical emittance, as in principle we can minimize their contribution by choosing weak bending magnets in them. On the other hand, we cannot choose as weak modulator as we want since it will also affect the energy modulation efficiency. This is the reason why the contribution of modulators to the vertical emittance is of more fundamental importance.

The resonant condition of the laser-electron interaction inside a planar undulator is

$$
\lambda_L = \frac{1 + \frac{K^2}{2}}{2\gamma^2} \lambda_u,
\tag{3.53}
$$

with $K = \frac{eB_0\lambda_u}{2\pi m_e c} = 0.934 \cdot B_0[T] \cdot \lambda_u[cm]$ the dimensionless undulator parameter. The effective modulation voltage of a laser modulator using a planar undulator is related to the laser and undulator parameters according to [19]

$$V_L = \frac{[JJ]K}{\gamma}\sqrt{\frac{4P_L Z_0 Z_R}{\lambda_L}}\tan^{-1}\left(\frac{L_u}{2Z_R}\right). \tag{3.54}$$

in which $[JJ] = J_0(\chi) - J_1(\chi)$ and $\chi = \frac{K^2}{4+2K^2}$, P_L is the modulation laser power, $Z_0 = 376.73\ \Omega$ is the impedance of free space, Z_R is the Rayleigh length of the modulation laser, L_u is the undulator length. The linear energy chirp strength around zero-crossing phase is therefore

$$h = \frac{eV_L}{E_0}k_L = \frac{e[JJ]K}{\gamma^2 mc^2}\sqrt{\frac{4P_L Z_0 Z_R}{\lambda_L}}\tan^{-1}\left(\frac{L_u}{2Z_R}\right)k_L, \tag{3.55}$$

where $k_L = 2\pi/\lambda_L$ is the wavenumber of the modulation laser.

For simplicity, we set $\epsilon_y = \Delta\epsilon_y(\text{Mod, QE})$, i.e., the vertical emittance is purely from the contribution of these two modulators, and assuming that equality holds in Theorem one, then the required modulation laser power is

$$P_L = \frac{\lambda_L}{4Z_0 Z_R}\left(\frac{\epsilon_y}{\sigma_{zy}(\text{Mod})\sigma_z(\text{Rad})\frac{e[JJ]K}{\gamma^2 mc^2}\tan^{-1}\left(\frac{L_u}{2Z_R}\right)k_L}\right)^2$$

$$= \frac{1}{([JJ]K)^2}\frac{\lambda_L^3}{3\pi^3 Z_0}\frac{55}{48\sqrt{3}}\frac{\alpha_F c^2 \lambda_e^2 \gamma^7 B_{0\text{Mod}}^3}{C_\gamma E_0^3 B_{\text{ring}}}\frac{1}{\sigma_z^2(\text{Rad})}\frac{\frac{L_u}{2Z_R}}{\left[\tan^{-1}\left(\frac{L_u}{2Z_R}\right)\right]^2}. \tag{3.56}$$

Now we try to derive more useful scaling laws to offer guidance in our parameters choice for a TLC SSMB ring. To maximize the energy modulation, we need $\frac{Z_R}{L_u} = 0.359 \approx \frac{1}{3}$. When $K > \sqrt{2}$, we approximate the resonance condition as $\lambda_L \approx \frac{K^2}{4\gamma^2}\lambda_u$, and $[JJ] \approx 0.7$. Then we have

$$P_L \propto \frac{\lambda_L^3}{K^2}\frac{\gamma^4 B_{0\text{Mod}}^3}{B_{\text{ring}}}\frac{1}{\sigma_z^2(\text{Rad})} \propto \lambda_L^{\frac{7}{3}}\gamma^{\frac{8}{3}}\frac{B_{0\text{Mod}}^{\frac{7}{3}}}{B_{\text{ring}}}\frac{1}{\sigma_z^2(\text{Rad})}. \tag{3.57}$$

The corresponding modulator length scaling is

$$L_u \propto \frac{B_{\text{ring}}\epsilon_y}{\mathcal{H}_y(\text{Mod})B_{0\text{Mod}}^3}. \tag{3.58}$$

Putting in the numbers for the constants, we obtain the quantitative expressions of the above scalings for practical use

$$P_L[\text{kW}] \approx 5.67 \frac{\lambda_L^{\frac{7}{3}}[\text{nm}] E_0^{\frac{8}{3}}[\text{GeV}] B_{0\text{Mod}}^{\frac{7}{3}}[\text{T}]}{\sigma_z^2(\text{Rad})[\text{nm}] B_{\text{ring}}[\text{T}]},$$

$$L_u[\text{m}] \approx 57 \frac{B_{\text{ring}}[\text{T}] \epsilon_y[\text{pm}]}{\mathcal{H}_y(\text{Mod})[\mu\text{m}] B_{0\text{Mod}}^3[\text{T}]}. \tag{3.59}$$

The above scaling laws are accurate when $K > \sqrt{2}$.

Note that ϵ_y does not appear explicitly in the scaling of the required laser power. It however affects the bunch length at the modulator and therefore the bunching factor at the radiator as we have explained. Also it affects the required modulator length. In other words, to obtain a desired bunching factor, the smaller ϵ_y is, the larger $\sigma_z(\text{Rad})$ we can use, thus a lower modulation laser power. Generally a shorter modulation laser wavelength and lower beam energy is preferred in lowering the required laser power. But we need to keep in mind that when the beam energy is too low, intrabeam scattering (IBS) could blow up ϵ_y [20, 21]. From the scaling, a weaker $B_{0\text{Mod}}$ means a smaller modulation laser power will be needed. But we should be aware that the corresponding length of modulator $L_u \propto \frac{1}{B_{0\text{Mod}}^3}$. In Table 6.2 of the final chapter, we have presented an example parameters set of a TLC SSMB storage ring for high-power EUV and soft X-ray radiation generation, based on the investigations presented here.

3.1.4 Transverse Deflecting RF or TEM01 Mode Laser for Coupling

Now we investigate in more detail about the application of TLC in SSMB for bunch compression, using a TEM01 mode laser modulator for the modulation kick. This belongs to the category of Theorem two.

3.1.4.1 TEM01 Mode Laser Modulator for Bunch Compression

A laser modulator implementing a TEM01 mode laser is like a transverse deflecting RF cavity in the optical wavelength range. The electric field of a Hermite-Gaussian TEM01 mode laser polarized in the vertical direction is [19]

$$E_y = E_{y0} e^{ikz - i\omega t} \left(\frac{1}{1 + i\frac{z}{Z_R}} \right)^2 \exp\left[i\frac{kQ}{2}(x^2 + y^2) \right] \frac{2\sqrt{2}}{w_0} y,$$

$$E_z = E_{y0} e^{ikz - i\omega t} \left(\frac{1}{1 + i\frac{z}{Z_R}} \right)^2 \exp\left[i\frac{kQ}{2}(x^2 + y^2) \right] \frac{2\sqrt{2}}{w_0} \left(\frac{i}{k} - Qy^2 \right), \tag{3.60}$$

with $Q = \dfrac{i}{Z_R\left(1+i\frac{z}{Z_R}\right)}$. The relation between E_{y0} and the laser peak power for a TEM01 mode laser is given by

$$P_L = \frac{E_{y0}^2 Z_R \lambda_L}{2Z_0}. \tag{3.61}$$

Note there is a factor of two difference in the above laser power formula compared to the case of a TEM00 mode laser. The electron wiggles in a vertical planar undulator according to

$$y(z) = \frac{K}{\gamma k_u} \sin(k_u z), \tag{3.62}$$

and the laser-electron exchanges energy according to

$$\frac{dW}{dt} = ev_y E_y + ev_z E_z. \tag{3.63}$$

Assuming that the laser beam waist is in the middle of the undulator, and when $x, y \ll w(z)$, which is typically the case in SSMB, we drop $\exp\left[i\frac{k_L Q}{2}(x^2 + y^2)\right]$ in the laser electric field. Since usually $Z_R \gg \lambda_u$, we can also drop the contribution from E_z on energy modulation. Then the integrated modulation voltage induced by the laser in the undulator is

$$V_{\text{mod}} = \frac{4K[JJ]}{\gamma} \frac{\sqrt{\pi P_L Z_0}}{\lambda_L} \frac{\frac{L_u}{2Z_R}}{1 + \left(\frac{L_u}{2Z_R}\right)^2} y, \tag{3.64}$$

The linear energy chirp with respect to y introduced is then

$$t = \frac{eV_{\text{mod}}}{E_0} \frac{1}{y} = \frac{2eK[JJ]k_L}{\gamma^2 mc^2} \sqrt{\frac{P_L Z_0}{\pi}} \frac{\frac{L_u}{2Z_R}}{1 + \left(\frac{L_u}{2Z_R}\right)^2}. \tag{3.65}$$

Note that the symplecticity of the dynamics requires that the vertical angle of the particle after modulation will depend on its initial longitudinal location. This observation is also supported by the Panofsky-Wenzel theorem [22]

$$\frac{\partial \Delta y'}{\partial s} = \frac{\partial}{\partial y_0}\left(\frac{\Delta \gamma}{\gamma}\right), \tag{3.66}$$

where $\Delta y'$ and $\Delta \gamma$ are the electron angular kick and energy change in the laser modulator. It is interesting to note that the modulation kick strength depends on the ratio between Z_R and L_u, instead their absolute values, and the maximal modulation is realized when $Z_R = \frac{L_u}{2}$.

Now we can do some evaluation based on the formulas. For an example choice of parameters, $E_0 = 400$ MeV, $\lambda_L = 270$ nm, $K[JJ] = 5$, $P_L = 1$ MW, $Z_R = \frac{L_u}{2}$, we have $t = 4$ m^{-1}. If $\epsilon_y = 1$ pm and the desired bunch length is 3 nm at the radiator, i.e., $\sigma_z(\text{Rad}) = 3$ nm, then according to Theorem two we have $\beta_y(\text{Mod}) \geq \frac{\epsilon_y}{t^2(\text{Mod})\sigma_z^2(\text{Rad})} = 6.9 \times 10^3$ m. The application of TLC with a TEM01 mode laser to compress the bunch length therefore faces the issue of a too large β_y at the modulator, if the desired bunch length is at nm level, which is needed for example in 13.5 nm coherent EUV radiation generation. However, if our target wavelength region is $\lambda_R \gtrsim 100$ nm, then the idea looks appealing as the required β_y is within the reasonable reach then. We remind the readers that there could be other more effective angular modulation scheme invented such that the issue of large β_y at the modulator can be solved, even if our target radiation wavelength is still in EUV.

3.1.4.2 Contribution of Modulators to Vertical Emittance

As said the advantage of TLC for bunch compression is based on a small vertical emittance. Like before let us now investigate the contribution of modulators to ϵ_y. We remind the readers that for bunch compression using a TEM01 mode laser, in principle, we can place the modulator at a dispersion-free location. Note however, if we aim at a complete y-z emittance exchange, the modulator needs to be placed at a dispersive location as will be discussed in next section. Here to minimize the contribution of modulators to ϵ_y, we choose to place the modulator at dispersion-free location, which means $d = 0$ and $d' = 0$, then the bunch compression condition is

$$\begin{aligned} T_{55} &= t R_{54} + 1 = 0, \\ T_{56} &= R_{56} = 0. \end{aligned} \tag{3.67}$$

Although we have placed the modulator at a dispersion-free location, there is still some residual contribution to ϵ_y since the transfer matrix of a TEM01 mode laser modulator is intrinsically transverse-longitudinal coupled, and the physical length L_u and $r_{56} = 2N_u\lambda_L$ of the modulator are nonzero. The thick-lens transfer matrix of the laser modulator can be obtained by slicing the laser modulator to tiny slices and use the thin-lens kick and drift method to get the total map. If we consider only terms to first order of t, r_{56} and L_u, then the thick-lens matrix of the TEM01 laser modulator is

$$\mathbf{M}_2 \approx \begin{pmatrix} 1 & L_u & \frac{tL_u}{2} & \frac{r_{56}tL_u}{6} \\ 0 & 1 & t & \frac{r_{56}t}{2} \\ \frac{r_{56}t}{2} & \frac{r_{56}tL_u}{6} & 1 & r_{56} \\ t & \frac{tL_u}{2} & 0 & 1 \end{pmatrix}. \tag{3.68}$$

Note that $\mathbf{M}_2$ is symplectic to first order of t. The transfer matrix of the state vector from the modulator entrance to a distance of s in it is

$$\mathbf{M}_{2s} = \begin{pmatrix} 1 & \left(\frac{s}{L_u}\right)L_u & \left(\frac{s}{L_u}\right)^2\frac{tL_u}{2} & \left(\frac{s}{L_u}\right)^3\frac{r_{56}tL_u}{6} \\ 0 & 1 & \left(\frac{s}{L_u}\right)t & \left(\frac{s}{L_u}\right)^2\frac{r_{56}t}{2} \\ \left(\frac{s}{L_u}\right)^2\frac{r_{56}t}{2} & \left(\frac{s}{L_u}\right)^3\frac{r_{56}tL_u}{6} & 1 & \left(\frac{s}{L_u}\right)r_{56} \\ \left(\frac{s}{L_u}\right)t & 0 & 0 & 1 \end{pmatrix}. \tag{3.69}$$

Here for simplicity we have assumed that the laser is a plane wave such that the induced angular modulation strength is proportional to the distance traveled inside the modulator.

Assuming that the one-turn map observed at the entrance of modulator is

$$\mathbf{T}(0) = \begin{pmatrix} \cos\Phi_y + \alpha_y\sin\Phi_y & \beta_y\sin\Phi_y & 0 & 0 \\ -\gamma_y\sin\Phi_y & \cos\Phi_y - \alpha_y\sin\Phi_y & 0 & 0 \\ 0 & 0 & \cos\Phi_z + \alpha_z\sin\Phi_z & \beta_z\sin\Phi_z \\ 0 & 0 & -\gamma_z\sin\Phi_z & \cos\Phi_z - \alpha_z\sin\Phi_z \end{pmatrix}, \tag{3.70}$$

in which $\Phi_y = 2\pi\nu_y$ and $\Phi_z = 2\pi\nu_s$. The eigenvector of the one-turn map corresponding to the vertical mode at the position s inside the modulator is then

$$\mathbf{E}_{II}(s) = \mathbf{M}_{2s}\mathbf{E}_{II}(0) = \frac{1}{\sqrt{2}}\begin{pmatrix} \sqrt{\beta_y} + \left(\frac{s}{L_u}\right)L_u\frac{i-\alpha_y}{\sqrt{\beta_y}} \\ \frac{i-\alpha_y}{\sqrt{\beta_y}} \\ \left(\frac{s}{L_u}\right)^2\frac{r_{56}t}{2}\sqrt{\beta_y} + \left(\frac{s}{L_u}\right)^3\frac{r_{56}tL_u}{6}\frac{i-\alpha_y}{\sqrt{\beta_y}} \\ \left(\frac{s}{L_u}\right)t\sqrt{\beta_y} \end{pmatrix}e^{i\Phi_{II}}. \tag{3.71}$$

Note that to ensure the TLC-based bunch compression can repeat turn-by-tun in a ring, usually two laser modulators are placed upstream and downstream of the radiator, respectively, to form a pair. According to Chao's SLIM formalism [23], we can calculate the contribution of the two modulators to ϵ_y

$$\Delta\epsilon_y(\text{Mod, QE}) = 2 \times \frac{55}{48\sqrt{3}}\frac{\alpha_F\lambda_e^2\gamma^5}{\alpha_V}\int_0^{L_u}\frac{|\mathbf{E}_{II5}(s)|^2}{|\rho(s)|^3}ds. \tag{3.72}$$

When $N_u \gg 1$, due to the fast oscillating behaviour of $\sin\left[2N_u\pi\left(\frac{s}{L_u}\right)\right]$, we can adopt the approximation

$$\Delta\epsilon_y(\text{Mod, QE}) \approx 2 \times \frac{55}{96\sqrt{3}}\frac{\alpha_F\lambda_e^2\gamma^5}{\alpha_V}\frac{1}{\rho_0^3}\frac{1}{L_u}\int_0^{L_u}\left|\sin\left[2N_u\pi\left(\frac{s}{L_u}\right)\right]\right|^3$$
$$\times \int_0^{L_u}\left|\left(\frac{s}{L_u}\right)^2\frac{r_{56}t}{2}\sqrt{\beta_y} + \left(\frac{s}{L_u}\right)^3\frac{r_{56}tL_u}{6}\frac{i-\alpha_y}{\sqrt{\beta_y}}\right|^2ds. \tag{3.73}$$

For simplicity, we assume $\alpha_y = 0$ at the modulator entrance, and usually $\beta_y(\text{Mod}) \gg L_u$, then

$$\Delta\epsilon_y(\text{Mod}, \text{QE}) \approx 2 \times \frac{55}{96\sqrt{3}} \frac{\alpha_F \lambda_e^2 \gamma^5}{\alpha_V} \frac{1}{\rho_{0\text{Mod}}^3} \frac{4}{3\pi} \frac{r_{56}^2 t^2 \beta_y}{20} L_u. \qquad (3.74)$$

Now let us put in some numbers to get a more concrete feeling. For example, if $E_0 = 400$ MeV, $\rho_{\text{ring}} = 1$ m ($B_{\text{ring}} = 1.33$ T), $\lambda_L = 270$ nm, $\lambda_u = 4$ cm, $K = 3.8$, $B_0 = 1.02$ T, $N_u = 10$, $r_{56} = 2N_u\lambda_L = 5.4$ μm, $L_u = 0.4$ m, $\epsilon_y = 1$ pm, $\beta_y(\text{Mod}) = 100$ m, $\sigma_y(\text{Mod}) = \sqrt{\epsilon_y\beta_y(\text{Mod})} = 10$ μm, $\sigma_z(\text{Rad}) = 2$ nm, $t = \frac{1}{\sqrt{\beta_y(\text{Mod})\mathcal{H}_y(\text{Rad})}} = \frac{\epsilon_y}{\sigma_y(\text{Mod})\sigma_z(\text{Rad})} = 50$ m^{-1}, then the contribution of the two modulators to ϵ_y is $\Delta\epsilon_y(\text{Mod}, \text{QE}) \approx 2.06$ fm. So generally, the contribution of the two modulators to ϵ_y is a small value, if the modulators are placed at dispersion-free locations.

3.1.5 Emittance Exchange

3.1.5.1 Lattice Condition

For completeness of the investigation, it might also be helpful to make a short discussion on the relation between our TLC analysis and the transverse-longitudinal emittance exchange (EEX). For a complete EEX, we need the transfer matrix of the form

$$\mathbf{T} = \begin{pmatrix} \mathbf{0} & \mathbf{B} \\ \mathbf{C} & \mathbf{0} \end{pmatrix}. \qquad (3.75)$$

Therefore, EEX is a special case in the context of our problem definition of TLC-based bunch compression, i.e., in EEX the final beam is also y-z decoupled. As can be seen from Eq. (3.31), the application of a normal RF or TEM00 mode laser modulator cannot accomplish a complete EEX, as $T_{65} = h \neq 0$. PEHG and ADM can thus be viewed as partial EEXs. In contrast, a transverse deflecting RF or TEM01 mode laser modulator can be used to obtain a complete EEX. All we need is to add another condition to Eq. (3.22), i.e.,

$$dt + 1 = 0. \qquad (3.76)$$

After some straightforward algebra, the relations in Eqs. (3.22) and (3.76) can be summarized in an elegant form as follows [24]

Fig. 3.7 Application of two transverse-longitudinal emittance exchangers to manipulate the bunch length in a storage ring

$$t = -\frac{1}{d},$$
$$D = R_{34}d' + R_{33}d,$$
$$D' = R_{44}d' + R_{43}d.$$

$$(3.77)$$

Note that the above relations mean that the lattices upstream and downstream the transverse deflecting RF are not mirror symmetric with respect to each other [25]. Under the conditions given in Eq. (3.77), we have

$$
\mathbf{T} = \begin{pmatrix}
0 & 0 & -\frac{R_{34}}{d}dR_{33} - R_{34}\frac{r_{56}-dd'}{d} \\
0 & 0 & -\frac{R_{44}}{d}dR_{43} - R_{44}\frac{r_{56}-dd'}{d} \\
dr_{43} - r_{33}\frac{R_{56}+dd'}{d} & dr_{44} - r_{34}\frac{R_{56}+dd'}{d} & 0 & 0 \\
-\frac{r_{33}}{d} & -\frac{r_{34}}{d} & 0 & 0
\end{pmatrix}
$$

$$
= \begin{pmatrix}
0 & 0 & T_{35} & T_{36} \\
0 & 0 & T_{45} & T_{46} \\
T_{53} & T_{54} & 0 & 0 \\
T_{63} & T_{64} & 0 & 0
\end{pmatrix}.
$$

$$(3.78)$$

3.1.5.2 Two EEXs as an Insertion

In order to apply EEX to generate short bunch in a storage ring on a turn-by-turn basis, another EEX might be needed following the radiator to swap back the ϵ_y and ϵ_z to maintain the ultrasmall vertical emittance ϵ_y. If there is only one transverse-longitudinal EEX, the ring will then be a transverse-longitudinal Möbius accelerator [26], which is also an interesting topic we are not going into in this dissertation.

Now we consider the application of two y-z EEXs for bunch length manipulation in a storage ring as shown in Fig. 3.7. The motivation is still to make use of the fact that the vertical emittance ϵ_y is rather small in a planar x-y uncoupled ring. The first natural idea is to add an inverse EEX unit following the EEX,

$$
\mathbf{T}^{-1} = \begin{pmatrix}
0 & 0 & T_{64} & -T_{54} \\
0 & 0 & -T_{63} & T_{53} \\
T_{46} & -T_{36} & 0 & 0 \\
-T_{45} & T_{35} & 0 & 0
\end{pmatrix},
$$

$$(3.79)$$

then the total insertion will be an identity matrix and be transparent to the ring. The issue of this approach, however, is that we need to design the downstream beamline with an R_{56} having opposite sign to the upstream beamline, which might be a challenging task if we aim at a compact lattice.

The second natural idea is to implement the mirror symmetry of the upstream beamline as the downstream beamline, which is straightforward for the lattice design. The transfer matrix of the mirror image is related to that of the original beamline according to [27, 28]

$$\mathbf{T}_{\text{mirror}}\mathbf{UT} = \mathbf{U}, \ \mathbf{U} = \begin{pmatrix} 1 & 0 & 0 & 0 \\ 0 & -1 & 0 & 0 \\ 0 & 0 & -1 & 0 \\ 0 & 0 & 0 & 1 \end{pmatrix}. \tag{3.80}$$

Therefore,

$$\mathbf{T}_{\text{mirror}} = \mathbf{UT}^{-1}\mathbf{U}^{-1} = \begin{pmatrix} 0 & 0 & -T_{64} & -T_{54} \\ 0 & 0 & -T_{63} & -T_{53} \\ -T_{46} & -T_{36} & 0 & 0 \\ -T_{45} & -T_{35} & 0 & 0 \end{pmatrix}. \tag{3.81}$$

Note, however, the transfer matrix of the total insertion in this case is generally not an identity matrix,

$$\mathbf{M} = \mathbf{T}_{\text{mirror}}\mathbf{T} = \begin{pmatrix} \mathbf{A'} & \mathbf{0} \\ \mathbf{0} & \mathbf{E'} \end{pmatrix}, \tag{3.82}$$

with

$$\mathbf{A'} = \begin{pmatrix} -(T_{53}T_{64} + T_{54}T_{63}) & -2T_{54}T_{64} \\ -2T_{53}T_{63} & -(T_{53}T_{64} + T_{54}T_{63}) \end{pmatrix},$$
$$\mathbf{E'} = \begin{pmatrix} -(T_{35}T_{46} + T_{36}T_{45}) & -2T_{36}T_{46} \\ -2T_{35}T_{45} & -(T_{35}T_{46} + T_{36}T_{45}) \end{pmatrix}. \tag{3.83}$$

A special case of EEX is the phase space exchange (PSX), i.e., the exchange happens in the phase space variables apart from a magnification factor. In this case, a PSX followed by its mirror can form an identity or a negative identity matrix. Case one:

$$\mathbf{T} = \begin{pmatrix} 0 & 0 & 0 & m_1 \\ 0 & 0 & -\frac{1}{m_1} & 0 \\ 0 & m_2 & 0 & 0 \\ -\frac{1}{m_2} & 0 & 0 & 0 \end{pmatrix}, \ \mathbf{T}_{\text{mirror}} = \begin{pmatrix} 0 & 0 & 0 & -m_2 \\ 0 & 0 & \frac{1}{m_2} & 0 \\ 0 & -m_1 & 0 & 0 \\ \frac{1}{m_1} & 0 & 0 & 0 \end{pmatrix}, \ \mathbf{M} = \mathbf{T}_{\text{mirror}}\mathbf{T} = \mathbf{I}, \tag{3.84}$$

Case two:

$$
\mathbf{T} = \begin{pmatrix} 0 & 0 & m_1 & 0 \\ 0 & 0 & 0 & \frac{1}{m_1} \\ m_2 & 0 & 0 & 0 \\ 0 & \frac{1}{m_2} & 0 & 0 \end{pmatrix}, \quad \mathbf{T}_{\text{mirror}} = \begin{pmatrix} 0 & 0 & -\frac{1}{m_2} & 0 \\ 0 & 0 & 0 & -m_2 \\ -\frac{1}{m_1} & 0 & 0 & 0 \\ 0 & -m_1 & 0 & 0 \end{pmatrix}, \quad \mathbf{M} = \mathbf{T}_{\text{mirror}}\mathbf{T} = -\mathbf{I}.
$$

(3.85)

According to Eq. (3.78), for case one, we need

$$
-\frac{R_{34}}{d} = 0,
$$

$$
d R_{43} + d' R_{44} - \frac{r_{56} R_{44}}{d} = 0,
$$

$$
d r_{43} - d' r_{33} - \frac{r_{33} R_{56}}{d} = 0,
$$

$$
-\frac{r_{34}}{d} = 0,
$$

(3.86)

and

$$
\mathbf{T} = \begin{pmatrix} 0 & 0 & 0 & \frac{d}{R_{44}} \\ 0 & 0 & -\frac{R_{44}}{d} & 0 \\ 0 & \frac{d}{r_{33}} & 0 & 0 \\ -\frac{r_{33}}{d} & 0 & 0 & 0 \end{pmatrix}.
$$

(3.87)

For case two, we need

$$
d R_{33} + d' R_{34} - \frac{r_{56} R_{34}}{d} = 0,
$$

$$
-\frac{R_{44}}{d} = 0,
$$

$$
d r_{44} - d' r_{34} - \frac{r_{34} R_{56}}{d} = 0,
$$

$$
-\frac{r_{33}}{d} = 0,
$$

(3.88)

and

$$
\mathbf{T} = \begin{pmatrix} 0 & 0 & -\frac{R_{34}}{d} & 0 \\ 0 & 0 & 0 & -\frac{d}{R_{34}} \\ -\frac{d}{r_{34}} & 0 & 0 & 0 \\ 0 & -\frac{r_{34}}{d} & 0 & 0 \end{pmatrix}.
$$

(3.89)

3.2   Nonlinear Transverse-Longitudinal Coupling Dynamics

3.2.1   Average Path Length Dependence on Betatron Amplitudes

After investigating the linear TLC, we will now examine nonlinear coupling. However, here we consider only the second-order path lengthening or shortening from betatron oscillations, and its impact on equilibrium beam parameters. A general discussion of the nonlinear dynamics is beyond the scope of this dissertation. The second-order TLC considered here originates from a dependence of the average path length on the betatron oscillation amplitudes, which can be expressed by a concise formula

$$\Delta C = -2\pi (\xi_x J_x + \xi_y J_y), \tag{3.90}$$

where ΔC is the average path-length deviation relative to the ideal particle, and $\xi_{x,y} = \frac{dv_{x,y}}{d\delta}$ and $J_{x,y}$ are the horizontal (vertical) chromaticity and betatron invariant, respectively. This simple relation is a result of the symplecticity of the Hamiltonian dynamics [29–31]. It is called a second-order coupling because the betatron invariant is a second-order term with respect to the transverse position and angle. Note that Eq. (3.90) is accurate only for the cases of multiple passes or multiple betatron oscillation periods as it is a betatron-phase-averaged result. For the case of a single pass with only a few betatron oscillations, there will be an extra term, depending on the betatron phase advance, in the path length formula.

This path length effect has previously been theoretically analyzed by several authors in different contexts [29–34]. Due to this effect, particles with different betatron amplitudes lose synchronization with each other when traversing a lattice with nonzero chromaticity. This leads to a stringent requirement on the beam emittance for FELs in the X-ray regime (XFELs), as microbunching can be smeared out by this effect when the beam is traveling through the undulator [35]. This effect is also crucial in non-scaling fixed-field alternating-gradient (FFAG) accelerators for muon acceleration [31], as a muon beam typically has a large emittance. Furthermore, the natural chromaticities of a linear non-scaling FFAG accelerator are usually not corrected to achieve a large transverse acceptance. This effect may also have an impact on the momentum and dynamic aperture in a storage ring [36, 37], for example, due to the Touschek scattering-induced large betatron amplitude or the large natural chromaticity in a low-emittance lattice. In this dissertation, we will emphasize the importance of this nonlinear TLC effect in precision longitudinal dynamics in a storage ring, such as SSMB.

3.2.2 Energy Widening and Distortion

As mentioned, this second-order TLC effect can disperse microbunching in XFELs. Methods of overcoming this influence are referred to as "beam conditioning". Several such methods have been proposed since the first publication of Ref. [35]. The basic idea of these proposals is to compensate the difference in path length through a difference in velocity by establishing a correlation between the betatron amplitude and the particle energy. In a storage ring, unlike in a single-pass device, the RF cavity will "condition" the beam automatically, causing all particles to synchronize with it in an average sense through phase stabilization (bunching). This is accomplished by introducing a betatron-amplitude-dependent energy shift to compensate for the path-length difference arising from the betatron oscillations,

$$\Delta\delta = -\frac{\Delta C}{\alpha C_0}, \tag{3.91}$$

where α is the momentum compaction factor of the ring, defined in Eq. (2.1). This shift will result in the beam energy widening in a quasi-isochronous ring with nonzero chromaticity, because different particles have different betatron invariants [34]. This widening will become more significant with the decreasing of the momentum compaction.

Due to the energy shift, there will also be an amplitude-dependent shift in the betatron oscillation center at dispersive locations. The shift direction is determined by the signs of α, $\xi_{x,y}$ and $D_{x,y}$, and the magnitude of the shift is determined by the magnitudes of $J_{x,y}$, α, $\xi_{x,y}$ and $D_{x,y}$. The physical pictures of the betatron center shift resulting from this effect are shown in Fig. 3.8.

When quantum excitation is also taken into account, the total relative energy deviation of a particle with respect to the ideal particle is

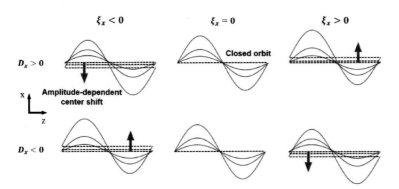

Fig. 3.8 Physical picture of the amplitude-dependent shift of betatron oscillation center in the case of a positive momentum compaction. Only a horizontal betatron oscillation is considered in this illustration

$$\delta = \Delta\delta + \delta_{\text{qe}}, \tag{3.92}$$

where δ_{qe} represents the quantum excitation contribution. Finding a general analytical formula for the steady-state distribution of the particles is a complex task and, at the same time, not very useful. Simpler expressions can be obtained based on reasonable approximations. Since the vertical emittance is usually much smaller than the horizontal emittance in a planar uncoupled ring, here, we consider only the contribution from the horizontal emittance. When the coupling is not very strong, the distributions of J_x and δ_{qe} are still approximately exponential and Gaussian, respectively, and are independent of each other,

$$\psi(J_x) = \frac{1}{2\pi\epsilon_{x0}} e^{-\frac{J_x}{\epsilon_{x0}}}, \quad \psi(\delta_{\text{qe}}) = \frac{1}{\sqrt{2\pi}\sigma_{\delta 0}} e^{-\frac{\delta_{\text{qe}}^2}{2\sigma_{\delta 0}^2}}, \tag{3.93}$$

where ϵ_{x0} and $\sigma_{\delta 0}$ are the natural horizontal emittance and energy spread. The distribution of δ is thus an exponentially modified Gaussian because it is the sum of an exponential and a normal random variable,

$$\psi(\delta) = \frac{|\lambda|}{2} e^{\frac{\lambda\left(\lambda\sigma_{\delta 0}^2 - 2\delta\right)}{2}} \operatorname{erfc}\left[\frac{\operatorname{sgn}(\lambda)\left(\lambda\sigma_{\delta 0}^2 - \delta\right)}{\sqrt{2}\sigma_{\delta 0}}\right], \tag{3.94}$$

where $\lambda = \frac{\alpha C_0}{2\pi\xi_x\epsilon_{x0}}$, $\operatorname{sgn}(x)$ is the sign function and $\operatorname{erfc}(x)$ is the complementary error function, defined as $\operatorname{erfc}(x) = 1 - \operatorname{erf}(x) = \frac{2}{\sqrt{\pi}}\int_x^\infty e^{-t^2}\,dt$. The direction of long non-Gaussian tail of the energy distribution is determined by the signs of α and ξ_x.

Because of the dispersion and dispersion angle, the non-Gaussian particle energy distribution can also be reflected in the transverse dimension. When this nonlinear coupling is considered, the horizontal position and angle of a particle in the storage ring are

$$x = \sqrt{2J_x\beta_x}\cos\varphi_x + D_x\left(\delta_{\text{qe}} + \frac{2\pi\xi_x J_x}{\alpha C_0}\right),$$
$$x' = -\sqrt{2J/\beta_x}(\alpha_x\cos\psi_x + \sin\psi_x) + D_x'\left(\delta_{\text{qe}} + \frac{2\pi\xi_x J_x}{\alpha C_0}\right). \tag{3.95}$$

It is assumed that the concept of the Courant-Snyder functions is still approximately valid in Eq. (3.95).

With these approximations, the variance of δ is then

$$\sigma_\delta^2 = \sigma_{\delta 0}^2 + \left(\frac{2\pi\epsilon_{x0}\xi_x}{\alpha C_0}\right)^2. \tag{3.96}$$

Table 3.1 Parameters of the MLS in the experiment

Parameter	Value	Description
C_0	48 m	Ring circumference
E_0	630 MeV	Beam energy
f_{rf}	500 MHz	RF frequency
V_{rf}	≤ 600 kV	RF voltage
U_0	9.1 keV	Radiation loss per turn
J_s	1.95	Longitudinal damping partition
τ_δ	11.4 ms	Longitudinal radiation damping time
ϵ_{x0}	250 nm	Horizontal emittance
$\sigma_{\delta 0}$	4.4×10^{-4}	Natural energy spread

By assuming the MLS parameters shown in Table 3.1 and applying $\alpha = 1 \times 10^{-4}$ and $\xi_x = 2$, one can find that the energy spread contributed by this effect can be as significant as its natural value.

As discussed in Ref. [34], a shift in the energy center corresponds to a shift in the synchronous RF phase ϕ_s,

$$\Delta\phi_s \approx J_s \tan\phi_s \Delta\delta, \tag{3.97}$$

where J_s is the longitudinal damping partition number and nominally $J_s \approx 2$. Therefore, particles with different betatron amplitudes will oscillate around different fixed points in the longitudinal dimension, thus lengthening the bunch. The change in the synchronous RF phase in a unit of the longitudinal coordinate, Δz_s, is related to the relative change in energy, $\Delta\delta$, according to

$$\frac{\Delta z_s}{\sigma_{z0}} = \frac{v_s J_s \tan\phi_s}{f_{RF}/f_{rev}|\alpha|}\frac{\Delta\delta}{\sigma_{\delta 0}} \propto \frac{1}{\sqrt{|\alpha|}}\frac{\Delta\delta}{\sigma_{\delta 0}}, \tag{3.98}$$

where v_s is the synchrotron tune, f_{rev} is the particle revolution frequency in the ring, σ_{z0} and $\sigma_{\delta 0}$ are the natural bunch length and energy spread, respectively.

The critical value of alpha, α_c, when the relative change of bunch length and energy spread are the same can be calculated to be

$$\frac{v_s J_s \tan\phi_s}{f_{RF}/f_{rev}|\alpha_c|} = 1 \Rightarrow |\alpha_c| = \frac{J_s T_0 \tan\phi_s}{\pi f_{RF}/f_{rev}\tau_\delta}, \tag{3.99}$$

where $\tau_\delta = 1/\alpha_L = \frac{2E_0}{J_s U_0}T_0$ is the longitudinal radiation damping time. As an example, we use the MLS parameters given in Table 3.1 and consider the application of an RF voltage of 500 kV, which corresponds to a synchronous RF phase of $\phi_s = 0.018$ rad. The critical value of alpha is then $|\alpha_c| \approx 2.1 \times 10^{-9}$, which is about four orders of magnitude smaller than the alpha value reachable at the present MLS.

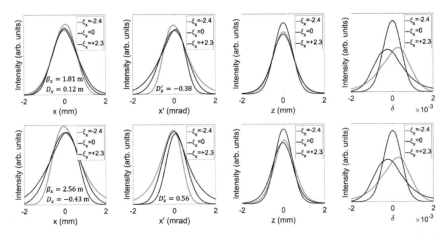

Fig. 3.9 Energy widening, bunch lengthening and distortion from a Gaussian distribution induced by a nonvanishing horizontal chromaticity. From up to bottom, the particle tracking results for distributions of x, x', z and δ are shown at two dispersive locations in the MLS under three different horizontal chromaticities ξ_x. The direction of the long non-Gaussian tail for δ is related to the signs of α and ξ_x, while for x and x' they are also dependent on D_x and D'_x, respectively. The simulation was conducted using the code ELEGANT [38] with a beam energy of 630 MeV, an RF voltage of 500 kV and the application of $\alpha = 1 \times 10^{-4}$. In each simulation, eight particles were tracked for 5×10^6 turns, corresponding to approximately 73 longitudinal radiation damping times

Therefore, the relative bunch lengthening resulting from this effect is much less significant than the corresponding energy widening at the MLS.

Several particle tracking simulations were conducted using the MLS lattice with the parameters presented in Table 3.1 to confirm the analysis. Two dispersive locations, with different signs and magnitudes of D_x, were selected as the observation points in the simulations. The simulation results are shown in Fig. 3.9. The energy widening and distortion from Gaussian behaviors are as expected and, indeed, are more significant than the bunch lengthening when $\alpha = 1 \times 10^{-4}$. At the two observation points, widening and distortion of the particle energy distribution also manifest in the transverse dimension through D_x and D'_x. The related optic functions at the two observation points are also shown in the profiles of x and x'. Note that the directions of the long non-Gaussian tails of the profiles and their relations to the signs of ξ_x, D_x and D'_x. We conclude that the simulation results agree well with the analysis and physical pictures presented above.

3.2.3 Experimental Verification

Here we report the first experimental verification of the energy widening and particle distribution distortion from Gaussian due to this second-order TLC effect as analyzed above. At the MLS, the Compton-backscattering (CBS) method is applied to measure

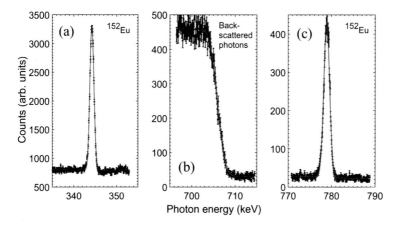

Fig. 3.10 Measurements of the CBS photons spectra at the 344.28 keV (**a**) and 778.90 keV (**c**) emission lines of ^{152}Eu radionuclide to calibrate the channel numbers in terms of keV and the HPGe-detector resolution at the CBS cutoff edge (**b**) in the CBS method of measuring beam energy spread

the electron energy [39]. Nevertheless, within certain limitations, the electron beam energy spread can also be evaluated from the CBS photon spectra [40]. The non-Gaussian momentum distribution makes the evaluation a bit more involved, but we can assume a Gaussian distribution with an equivalent mean energy spread. This is a good approximation as long as ξ_x is not too large.

The experiment is conducted with all 80 RF buckets equally filled. To exclude a severe impact from energy widening collective effects, the beam current is decayed till the horizontal beam size is not sensitively dependent on it. The average single-bunch current is below 12.5 μA (1 electron/1 pA) while doing the CBS measurements. To mitigate the influence of a nonlinear momentum compaction, the longitudinal chromaticity has been corrected close to zero. The other parameters of the ring in the experiment are presented in Table 3.1.

To get the energy spread based on the CBS method with precision, the HPGe-detector used in the measurement should be calibrated in terms of the photon energy per channel. This is realized by recording the emission lines from a ^{152}Eu radionuclide simultaneously during the measurement of the CBS photons. Moreover, the width of the fitted ^{152}Eu lines that are close to the edge of the CBS photons have been used to determine the detector resolution σ_{det} at the photon energy of the CBS cutoff edge, E_{edge}, in our case 707 keV. This is done by a linear interpolation of the width of the ^{152}Eu lines at 344.28 and 778.90 keV. The detector resolution σ_{det} at 707 keV is thus determined to be 0.64(4) keV. The calibration scheme and result is shown in Fig. 3.10.

Figure 3.11a shows the typical CBS photon spectra close to the cutoff edge under the cases of different ξ_x and α. The adjustment of ξ_x is accomplished by the implementation of different chromatic sextupole strengths and the α by slightly tuning

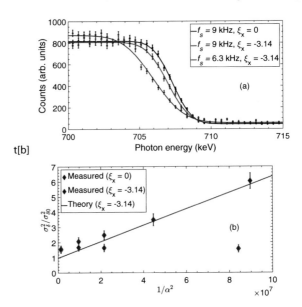

t[b]

Fig. 3.11 Measurement of the electron beam energy widening brought by the horizontal chromaticity using the CBS method. **a** The cutoff edges of CBS photon spectra under different ξ_x and f_s (therefore α). **b** Quantitative evaluation of the cut-off edges revealing the energy spread and its comparison with theory Eq. (3.96). The error bars in both figures are the one sigma uncertainties of the measurements and are due to calibration errors and counting statistics. The data acquisition of each spectrum takes 15 min

quadrupole strengths. In the experiment the RF voltage is kept constant and the synchrotron frequency f_s is proportional to the square root of the magnitude of α. The edge in the figure is a convolution of a step function representing the CBS cutoff edge with a Gaussian function which attributes to the finite HPGe-detector energy resolution and the electron beam energy spread. The fitted line is basically an error function from which the energy width of the CBS photons at the edge σ_{edge}, and therefore the electron beam energy spread σ_δ, can be deduced. It is assumed σ_{edge} is given by $\sigma_{edge} = \sqrt{\sigma_{det}^2 + (2E_{edge}\sigma_\delta)^2}$. The second term in the square root is due to the electron beam energy spread and is based on the fact that the energy of the backscattered photon is proportional to the electron energy squared.

It can be seen from Fig. 3.11a that the edge slope decreases with the magnitude increasing of ξ_x and lowering of α when $\xi_x \neq 0$, which indicates that there is an energy widening in the process. Quantitative evaluation of the edges revealing the energy spread and its comparison with theory of Eq. (3.96) are shown in Fig. 3.11b. The energy spread grows significantly with the magnitude decrease of α when $\xi_x = -3.14$ while it stays almost constant in the case of $\xi_x = 0$. The agreement between measurements and theory is quite satisfactory. This is the first direct experimental proof of the impact of this effect on the equilibrium beam parameters in a storage ring.

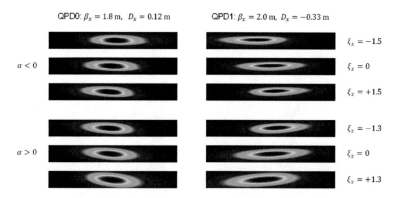

Fig. 3.12 Transverse beam intensity distortion from Gaussian at dispersive locations due to a non-vanishing horizontal chromaticity measured by the imaging systems installed at QPD0 and QPD1. Three different ξ_x are applied in both the positive and negative momentum compaction modes with $\alpha = \pm 8.4 \times 10^{-5}$. There is some residual horizontal-vertical coupling in the positive momentum compaction case, which do not influence the principle observation of the non-Gaussian behavior

As analyzed before, the bunch lengthening due to this second-order coupling is much less notable and also due to the limited resolution of the present streak camera, we do not measure the bunch lengthening in the experiment. Nevertheless, a more comprehensive investigation of this effect can be conducted on the other beam characteristics like the transverse intensity distribution. As can be seen from Eq. (3.95), particles with different betatron amplitudes oscillate around different closed orbits, which is the amplitude dependent center shift [41]. Because of the dispersion, the non-Gaussian particle momentum distribution can also be reflected to the transverse dimension, which can be observed by the beam imaging systems installed at the MLS [42].

Figure 3.12 shows the typical transverse beam intensity distribution measured by the imaging systems at two dispersive locations, QPD0 and QPD1, with different values of ξ_x in both the negative and positive momentum compaction modes. The relevant optics functions, β_x and D_x, at the two observation points are also shown in the figure. Note that D_x have different signs and magnitudes at QPD0 and QPD1. It can be seen that the horizontal beam distribution at these dispersive locations becomes asymmetric when $\xi_x \neq 0$. The long tail direction and the magnitude of deviation from Gaussian are determined by the signs and magnitudes of α, D_x, ξ_x and also the value of ϵ_{x0} and β_x, which fits with the expectations.

Figure 3.13a demonstrates the typical horizontal beam profiles measured at QPD1 in the negative momentum compaction mode under three different ξ_x and their good agreements with theory. It turns out that both the theoretical and experimental measured horizontal coordinate distribution $\psi(x)$ can be excellently fitted by a skewed Gaussian function

$$\psi(x) = \frac{1}{\sqrt{2\pi}\sigma} \cdot e^{-\frac{(x-b)^2}{2\sigma^2}} \cdot \left(1 + \mathrm{erf}\left[d \cdot \frac{x-b}{\sqrt{2}\sigma}\right]\right). \qquad (3.100)$$

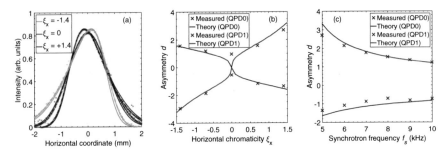

Fig. 3.13 Horizontal beam profile distortion from Gaussian by horizontal chromaticty. **a** Typical horizontal beam profile at QPD1 with $\alpha = -7 \times 10^{-5}$ under three different ξ_x. The closed orbit movements of the ideal particle due to the sextupole strengths changes when adjusting ξ_x have been compensated in the plot. Cross: beam imaging system measurement results. Dashed line: fit of the measurement data by an exponentially modified Gaussian function Eq. (3.100). Solid line: theoretical prediction. **b** Measured and theoretical asymmetry parameter d versus ξ_x at QPD0 and QPD1 with $f_s = 5$ kHz ($\alpha = -7 \times 10^{-5}$); **c** Measured and theoretical asymmetry parameter d versus f_s at QPD0 and QPD1 with $\xi_x = 1.4$. All the theoretical curves are obtained based on Eqs. (3.93) and (3.95)

The asymmetry parameter d in Eq. (3.100) is used to quantitatively describe the deviation from Gaussian and as a criterion to do comparison between measurements and theory. Figure 3.13b and c show the asymmetry parameter d versus ξ_x and f_s, therefore α, from measurements and theory at QPD0 and QPD1. It can be seen that the larger the ξ_x and the smaller the α, the more asymmetric the distribution is. Also the asymmetry at QPD1 is more significant than that at QPD0 as the magnitude of D_x at QPD1 is larger while the β_x difference at two places is not much. The agreement between measurements and theory confirms that this effect distorts the beam from Gaussian in both the longitudinal and transverse dimensions.

While the energy widening and beam distortion could be a detrimental outcome for some applications, it may actually also be beneficial as it can help to stabilize collective instabilities. The bunch lengthening on the other hand is much less notable compared to the energy widening. So quasi-isochronous ring-based coherent radiation schemes, like some of the SSMB scenarios, may boost the stable coherent radiation power by taking advantage of this effect. For example, the stable single-bunch current at the MLS can grow for more than one order of magnitude by increasing the absolute value of the horizontal chromaticity from zero to a value larger than three, with the head-tail and the other collective effects like the longitudinal microwave instability properly suppressed. It has been proved at the MLS that the increase of THz power due to a higher stable beam current overcompensates the decrease due to the slight bunch lengthening of this effect. Therefore, this is now the standard low momentum compaction mode at the MLS for the application of Fourier Transform Spectroscopy.

This nonlinear TLC may also be useful in some more applications. For example, it can be used for the real-time emittance evaluation in storage rings if the chromaticities, beta function and dispersion are known, which are usually easier to get than measuring the emittance directly. The amplitude dependent center shift can be applied to detect beam instabilities which blow up the transverse emittance [43]. A strongly asymmetric particle momentum distribution due to this effect cooperating with a large momentum compaction lattice can generate a strongly asymmetric distributed current, which is favored in some applications such as beam-driven wakefield acceleration [44].

References

1. Deng XJ, Chao AW, Feikes J, Huang WH, Ries M, Tang CX (2020) Single-particle dynamics of microbunching. Phys Rev Accel Beams 23:044002
2. Deng X, Huang W, Li Z, Tang C (2021) Harmonic generation and bunch compression based on transverse-longitudinal coupling. Nucl Instrum Methods Phys Res A 1019:165859
3. Li Z, Deng X, Pan Z, Tang C, Chao A (2023) A generalized longitudinal strong focusing storage ring, Generalized longitudinal strong focusing in a steady-state microbunching storage ring, submitted
4. Deng XJ, Klein R, Chao AW, Hoehl A, Huang WH, Li J, Lubeck J, Petenev Y, Ries M, Seiler I, Tang CX, Feikes J (2020) Widening and distortion of the particle energy distribution by chromaticity in quasi-isochronous rings. Phys Rev Accel Beams 23:044001
5. Shoji Y (2004) Bunch lengthening by a betatron motion in quasi-isochronous storage rings. Phys Rev ST Accel Beams 7:090703
6. Wüstefeld G (2005) Horizontal-longitudinal coupling in the fel beam line, Technical Report, BESSY TB 227/05
7. Huang X (2007) Matrix formalism of synchrobetatron coupling. Phys Rev ST Accel Beams 10:014002
8. Shimada M, Katoh M, Adachi M, Tanikawa T, Kimura S, Hosaka M, Yamamoto N, Takashima Y, Takahashi T (2009) Transverse-longitudinal coupling effect in laser bunch slicing. Phys Rev Lett 103:144802
9. Shoji Y (2011) Transient bunch lengthening by a betatron motion along bending sections. Nucl Instrum Methods Phys Res, Sect A 637(1):S120–S122
10. Courant ED, Snyder HS (1958) Theory of the alternating-gradient synchrotron. Ann Phys (NY) 3(1):1–48
11. Deng X, Chao A, Feikes J, Hoehl A, Huang W, Klein R, Kruschinski A, Li J, Matveenko A, Petenev Y et al (2021) Experimental demonstration of the mechanism of steady-state microbunching. Nature 590(7847):576–579
12. Kruschinski A, Feikes J, Li J, Ries M, Deng X, Hoehl A, Klein R (2023) Exploring the necessary conditions for steady-state microbunching at the metrology light source. In: Proceedings of the 14th international particle accelerator conference (IPAC'23), Vinece, Italy, 2023. JACoW, Geneva, p MOPA176
13. Deng H, Feng C (2013) Using off-resonance laser modulation for beam-energy-spread cooling in generation of short-wavelength radiation. Phys Rev Lett 111:084801
14. Feng C, Deng H, Wang D, Zhao Z (2014) Phase-merging enhanced harmonic generation free-electron laser. New J Phys 16(4):043021
15. Feng C, Zhao Z (2017) A storage ring based free-electron laser for generating ultrashort coherent euv and x-ray radiation. Sci Rep 7(1):4724
16. Dragt AJ (2021) Lie methods for nonlinear dynamics with applications to accelerator physics
17. Ratner D, Chao A (2011) Seeded radiation sources with sawtooth waveforms. In: Proceedings of the 33th international free electron laser conference (FEL'11), Shanghai, China, 2011, pp 53–56

18. Stupakov G, Zolotorev M (2011) Using laser harmonics to increase bunching factor in eehg. In: Proceedings of the 33th international free electron laser conference (FEL'11), Shanghai, China, 2011, pp 45–48
19. Chao A (2022) Focused laser, unpublished note
20. Piwinski A (1974) Intra-beam-scattering. In: Proceedings of the 9th international conference on high energy accelerators, Stanford, p 405
21. Bjorken JD, Mtingwa SK (1982) Intrabeam scattering. Part Accel 13(FERMILAB-PUB-82-47-THY):115–143
22. Panofsky W, Wenzel W (1956) Some considerations concerning the transverse deflection of charged particles in radio-frequency fields. Rev Sci Instrum 27(11):967–967
23. Chao AW (1979) Evaluation of beam distribution parameters in an electron storage ring. J Appl Phys 50(2):595–598
24. Fliller R (2006) General solution to a transverse to longitudinal emittance exchange beamline. Fermi Beams Document 2553–v2
25. Xiang D, Chao A (2011) Emittance and phase space exchange for advanced beam manipulation and diagnostics. Phys Rev ST Accel Beams 14:114001
26. Talman R (1995) A proposed möbius accelerator. Phys Rev Lett 74:1590–1593
27. Berz M, Makino K, Wan W (2015) An introduction to beam physics. CRC Press
28. Chao AW (2020) Lectures on accelerator physics. World Scientific
29. Forest É (1998) Beam dynamics: a new attitude and framework. Harwood Academic Publisher, Amsterdam
30. Chao AW (2022) Special topics in accelerator physics. World Scientific
31. Berg JS (2007) Amplitude dependence of time of flight and its connection to chromaticity. Nucl Instrum Methods Phys Res, Sect A 570(1):15–21
32. Artamonov A, Derbenev IS, Inozemtsev N (1989) The possibility of controlling the dispersion characteristics of particle motion in an undulator. Zh Tekh Fiz 59:214–216
33. Emery L (1993) Coupling of betatron motion to the longitudinal plane through path lengthening in low-α_c storage rings. In: Rossbach J (ed.) Proceedings of the 15th international conference on high energy accelerators. World Scientific, Singapore, pp 1172–1174
34. Shoji Y (2005) Dependence of average path length betatron motion in a storage ring. Phys Rev ST Accel Beams 8:094001
35. Sessler AM, Whittum DH, Yu L-H (1992) Radio-frequency beam conditioner for fast-wave free-electron generators of coherent radiation. Phys Rev Lett 68:309–312
36. Takao M (2008) Impact of betatron motion on path lengthening and momentum aperture in a storage ring. In: Proceedings of the 11th European particle accelerator conference (EPAC'08), Genoa, Italy, 2008, pp 3152–3154
37. Hoummi L, Lopez JR, Welsch C, Loulergue A, Nagaoka R (2019) Beam dynamics in mba lattices with different chromaticity correction schemes. In: Proceedings of the 10th international particle accelerator conference (IPAC'19), Melbourne, Australia, 2019. JACoW, Geneva, pp 346–349
38. Borland M (2000) Elegant: a flexible sdds-compliant code for accelerator simulation. Technical report, Argonne National Lab
39. Klein R, Brandt G, Fliegauf R, Hoehl A, Müller R, Thornagel R, Ulm G, Abo-Bakr M, Feikes J, Hartrott MV, Holldack K, Wüstefeld G (2008) Operation of the metrology light source as a primary radiation source standard. Phys Rev ST Accel Beams 11:110701
40. Klein R, Mayer T, Kuske P, Thornagel R, Ulm G (1997) Beam diagnostics at the bessy i electron storage ring with compton backscattered laser photons: measurement of the electron energy and related quantities. Nucl Instrum Methods Phys Res A 384(2–3):293–298
41. Shoji Y, Takao M, Nakamura T (2014) Amplitude dependent shift of betatron oscillation center. Phys Rev ST Accel Beams 17:064001
42. Koschitzki C, Hoehl A, Klein R, Thornagel R, Feikes J, Hartrott M, Wüstefeld G (2010) Highly sensitive beam size monitor for pa currents at the mls electron storage ring. In: Proceedings of the 1st international particle accelerator conference (IPAC'10), Kyoto, Japan, 2010. JACoW, Geneva, pp 894–896

43. Kobayashi K, Schimizu J, Hara T, Seike T, Soutome K, Takao M, Nakamura T (2011) Amplitude dependent betatron oscillation center shift by non-linearity and beam instability interlock. Conf Proc 110904:2178–2180
44. Chen P, Su JJ, Dawson JM, Bane KLF, Wilson PB (1986) Energy transfer in the plasma wake-field accelerator. Phys Rev Lett 56:1252–1255

Chapter 4
SSMB Radiation

Having discussed the methods to form and preserve microbunching in the last two chapters, now we present the theoretical and numerical study of the average and statistical property of coherent radiation from SSMB. Our results show that a kW-level average power quasi-continuous-wave EUV radiation can be obtained from an SSMB ring, provided that an average current of 1 A and bunch length of 3 nm microbunch train can be formed at the radiator which is assumed to be an undulator. Together with its narrowband feature, the EUV photon flux can reach $10^{15} \sim 10^{16}$ phs/s within a 0.1 meV energy bandwidth, which is three orders of magnitude higher than that in a conventional synchrotron source, allowing sub-meV resolution in angle-resolved photoemission spectroscopy (ARPES) and providing new opportunities for fundamental physics research. In the theoretical investigation, we have generalized the definition and derivation of the transverse form factor of an electron beam which can quantify the impact of its transverse size on coherent radiation. In particular, we have shown that the narrowband feature of SSMB radiation is strongly correlated with the finite transverse electron beam size. Considering the pointlike nature of electrons and quantum nature of radiation, the coherent radiation fluctuates from microbunch to microbunch, or for a single microbunch from turn to turn. Some important results concerning the statistical property of SSMB radiation are presented, with a brief discussion on its potential applications for example the beam diagnostics. The presented work is of value for the development of SSMB and better serve the potential synchrotron radiation users. In addition, it also sheds light on understanding the radiation characteristics of free-electron lasers (FELs), coherent harmonic generation (CHG), etc. Parts of the work presented in this chapter have been published in Ref. [1].

© The Author(s) 2024
X. Deng, *Theoretical and Experimental Studies on Steady-State Microbunching*,
Springer Theses, https://doi.org/10.1007/978-981-99-5800-9_4

4.1 Formulation of Radiation from a 3D Rigid Beam

For simplicity, as the first step we consider only the impacts of particle position x, y and z, but ignore the particle angular divergence x', y' and energy deviation δ, on the radiation. Under this approximation, concise and useful analytical formulas of the coherent radiation can be obtained. This approximation is accurate when the transverse and longitudinal beam size do not change much inside the radiator, i.e., $\beta_{x,y} \gtrsim L_r$ and $\beta_z \gtrsim R_{56,r}$, where $\beta_{x,y,z}$ are the Courant-Snyder functions of the beam in the horizontal, vertical and longitudinal dimensions [2], L_r and $R_{56,r}$ are the length and momentum compaction of the radiator, respectively. Here in this dissertation, we call this approximation the three-dimensional (3D) rigid beam approximation, as the beam sizes do not change much during radiation. We will see later in Sect. 4.6 that the conditions of rigid beam approximation is generally satisfied in the envisioned EUV SSMB. In addition, we will briefly discuss the impact of beam divergence and energy spread on coherent radiation in Sect. 4.4.

Assuming that the vector potential of radiation from the reference particle at the observation location is $\mathbf{A}_{\text{point}}(\theta, \varphi, t)$, with θ and φ being the polar and azimuthal angles in a spherical coordinate system, respectively, as shown in Fig. 4.1. Under far-field approximation, the vector potential of radiation from a 3D rigid electron beam containing N_e electrons is then

Fig. 4.1 Coordinate system used to calculate the undulator radiation spectrum. The magnetic field is in the y-direction, and the electron wiggles in the x-z plane

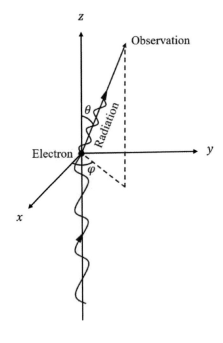

$$
\mathbf{A}_{\text{beam}}(\theta, \varphi, t) = N_e \int_{-\infty}^{\infty} \int_{-\infty}^{\infty} \int_{-\infty}^{\infty}
$$

$$
\mathbf{A}_{\text{point}}\left(\theta, \varphi, t + \frac{x \sin\theta \cos\varphi + y \sin\theta \sin\varphi}{c} + \frac{z}{\beta c}\right) \rho(x, y, z) dx dy dz, \tag{4.1}
$$

in which β is the particle velocity normalized by the speed of light in vacuum c, and $\rho(x, y, z)$ is the normalized charge density satisfying $\int_{-\infty}^{\infty} \int_{-\infty}^{\infty} \int_{-\infty}^{\infty} \rho(x, y, z)$ $dx dy dz = 1$. Note that we have assumed that the particle motion pattern, therefore also the radiation pattern of a single electron, does not depend on x, y and z of the particle. In other words, x, y, z of a particle influences only the arrival time of the radiation at the observation. This is the reason why their impacts can be treated within a single framework. The impacts of x', y' and δ are different. Generally, their impacts are two-fold. First, they affect the radiation of the single particle itself, i.e., the radiation pattern. Second, they affect the electron beam distribution, therefore the coherence of different particles, during the radiation process.

According to the convolution theorem, for a 3D rigid beam, we now have

$$
\mathbf{A}_{\text{beam}}(\theta, \varphi, \omega) = N_e \mathbf{A}_{\text{point}}(\theta, \varphi, \omega) b(\theta, \varphi, \omega), \tag{4.2}
$$

where

$$
b(\theta, \varphi, \omega) = \int_{-\infty}^{\infty} \int_{-\infty}^{\infty} \int_{-\infty}^{\infty} \rho(x, y, z) e^{-i\omega\left(\frac{x \sin\theta \cos\varphi + y \sin\theta \sin\varphi}{c} + \frac{z}{\beta c}\right)} dx dy dz. \tag{4.3}
$$

Since $\mathbf{A}(\theta, \varphi, t)$ is real, then $\mathbf{A}(\theta, \varphi, -\omega) = \mathbf{A}^*(\theta, \varphi, \omega)$. The energy radiated per unit solid angle per unit frequency interval is [3]

$$
\frac{d^2 W}{d\omega d\Omega}(\theta, \varphi, \omega) = 2|\mathbf{A}(\theta, \varphi, \omega)|^2. \tag{4.4}
$$

Therefore, we have

$$
\left.\frac{d^2 W}{d\omega d\Omega}(\theta, \varphi, \omega)\right|_{\text{beam}} = N_e^2 |b(\theta, \varphi, \omega)|^2 \left.\frac{d^2 W}{d\omega d\Omega}(\theta, \varphi, \omega)\right|_{\text{point}}. \tag{4.5}
$$

The total radiation energy spectrum of a beam can be obtained by the integration with respect to the solid angle

$$
\left.\frac{dW}{d\omega}\right|_{\text{beam}} = \int_0^\pi \sin\theta d\theta \int_0^{2\pi} d\varphi \left.\frac{d^2 W}{d\omega d\Omega}(\theta, \varphi, \omega)\right|_{\text{beam}}, \tag{4.6}
$$

and the total radiation energy of the beam can be calculated by further integration with respect to the frequency

$$W_{\text{beam}} = \int_0^{+\infty} d\omega \frac{dW}{d\omega}\bigg|_{\text{beam}}. \tag{4.7}$$

The reason why the lower limit in the above integration is 0, instead of $-\infty$, is that there is a factor of 2 in the right hand side of Eq. (4.4). The above formulas can be used to numerically calculate the characteristics of radiation from an electron beam, once its 3D distribution is given. Note that in the relativistic case, we only need to account for θ several times of $\frac{1}{\gamma}$, as the radiation is very collimated in the forward direction.

4.2 Form Factors

When the longitudinal and transverse dimensions of the electron beam are decoupled, we can factorize $b(\theta, \varphi, \omega)$ as

$$b(\theta, \varphi, \omega) = b_\perp(\theta, \varphi, \omega) \times b_z(\omega), \tag{4.8}$$

where

$$b_\perp(\theta, \varphi, \omega) = \int_{-\infty}^{\infty} \int_{-\infty}^{\infty} \rho(x, y) e^{-i\omega\left(\frac{x \sin\theta \cos\varphi + y \sin\theta \sin\varphi}{c}\right)} dx dy, \tag{4.9}$$

and

$$b_z(\omega) = \int_{-\infty}^{\infty} \rho(z) e^{-i\omega\frac{z}{\beta c}} dz. \tag{4.10}$$

Note that $\rho(x, y)$ and $\rho(z)$ are then the projected charge density. $b_z(\omega)$ is the usual bunching factor found in literature and is independent of the observation angle. This however is not true for $b_\perp(\theta, \varphi, \omega)$. For example, in the case of a 3D Gaussian x-y-z decoupled beam,

$$|b_\perp(\theta, \varphi, \omega)|^2 = \exp\left\{-\left(\frac{\omega}{c}\right)^2 \left[(\sigma_x \sin\theta \cos\varphi)^2 + (\sigma_y \sin\theta \sin\varphi)^2\right]\right\},$$
$$|b_z(\omega)|^2 = \exp\left[-\left(\frac{\omega}{\beta c}\right)^2 \sigma_z^2\right], \tag{4.11}$$

where $\sigma_{x,y,z}$ are the root-mean-square (RMS) size of the beam in the horizontal, vertical and longitudinal dimension, respectively.

In order to efficiently quantify the impact of the transverse and longitudinal distributions of an electron beam on the overall radiation energy spectrum, here we define the transverse and longitudinal form factors of an electron beam as

$$FF_\perp(\omega) = \frac{\int_0^\pi \sin\theta d\theta \int_0^{2\pi} d\varphi |b_\perp(\theta,\varphi,\omega)|^2 \frac{d^2W}{d\omega d\Omega}(\theta,\varphi,\omega)\Big|_{\text{point}}}{\int_0^\pi \sin\theta d\theta \int_0^{2\pi} d\varphi \frac{d^2W}{d\omega d\Omega}(\theta,\varphi,\omega)\Big|_{\text{point}}}, \tag{4.12}$$

and

$$FF_z(\omega) = |b_z(\omega)|^2, \tag{4.13}$$

respectively. The overall form factor is then

$$FF(\omega) = FF_\perp(\omega)FF_z(\omega). \tag{4.14}$$

The total radiation energy spectrum of a beam is related to that of a single electron by

$$\frac{dW}{d\omega}\Big|_{\text{beam}} = N_e^2 FF(\omega)\frac{dW}{d\omega}\Big|_{\text{point}}. \tag{4.15}$$

4.2.1 Longitudinal Form Factor

The longitudinal form factor is the usual bunching factor squared, and have been discussed extensively in literature. For example, the longitudinal form factor for a single Gaussian microbunch is given Eq. (4.11). When there are multi microbunches separated with each other a distance of the modulation laser wavelength λ_L like that in SSMB, the longitudinal form factor is that of the single bunch multiplied with a macro form factor,

$$FF_{z\text{MB}}(\omega) = FF_{z\text{SB}}(\omega)\left(\frac{\sin\left(N_b \frac{\omega}{c}\frac{\lambda_L}{2}\right)}{N_b \sin\left(\frac{\omega}{c}\frac{\lambda_L}{2}\right)}\right)^2, \tag{4.16}$$

where the subscripts $_{\text{MB}}$ and $_{\text{SB}}$ mean multi bunch and single bunch, respectively, and N_b is the number of microbunches. This macro form factor of multi bunches is a periodic function of the radiation frequency, with a period of the modulation laser frequency. The full width at half maximum (FWHM) linewidth around each laser harmonics is $\Delta\omega_{\text{FWHM}} = \frac{\omega_L}{N_b}$. When N_b goes to infinity, this macro form factor will become the periodic delta function. Figure 4.2 presents an example plot of the macro form factor for three different N_b.

When the radiation wavelength is a high harmonic of the modulation laser wavelength, corresponding to these delta function lines in the longitudinal form factor or radiation energy spectrum, there will be interference rings in the spatial distribution of the coherent radiation from different microbunches. The polar angles of these rings, corresponding to the delta function lines in energy spectrum, are determined

Fig. 4.2 Macro form factor of multi bunches, as a function of the number of bunches

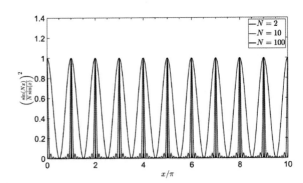

by the off-axis resonant condition. Note that when we use the above form factor $FF_{zMB}(\omega)$ to calculate the radiation energy or spectrum from N_b microbunches, the number of electrons used should be $N_b N_{eSB}$, with N_{eSB} the number of electrons per microbunch. We remind the readers that the electron beam energy spread and angular divergence will make the linewidth of the delta function lines analyzed above become non-zero. For example, the relative bandwidth of the radiation caused by an energy spread of σ_δ is $2\sigma_\delta$.

To make our results more useful, here we also present some analysis applies for FEL and CHG. As analyzed in Sect. 2.2.1, the bunching factor for a coasting beam-based CHG is

$$b_{z,\text{coasting}}(\omega) = \sum_{n=0}^{\infty} \delta\left(\frac{\omega}{c} - nk_L\right) J_n\left[-\frac{\omega}{c}R_{56}A\right] \exp\left[-\frac{1}{2}\left(\frac{\omega}{c}R_{56}\sigma_\delta\right)^2\right],$$
(4.17)

where

$$\delta(x) = \begin{cases} 1, & x = 0, \\ 0, & \text{else.} \end{cases}$$
(4.18)

Let us now consider the more-often confronted case of a finite bunch length. We assume that the initial beam current before microbunching is Gaussian with an RMS bunch length of σ_z. According to the convolution theorem, then

$$b_{z,\text{bunched}}(\omega) = b_{z,\text{coasting}}(\omega) \otimes b_{z,\text{Gaussian}}(\omega),$$
(4.19)

where $\otimes$ means convolution and $b_{z,\text{Gaussian}}(\omega) = \exp\left[-\frac{1}{2}\left(\frac{\omega}{c}\sigma_z\right)^2\right]$. Therefore, the longitudinal form factor of the beam is now $FF_{z,\text{bunched}}(\omega) = |b_{z,\text{bunched}}(\omega)|^2$. The convolution with $b_{z,\text{Gaussian}}(\omega)$ results in a non-zero bandwidth of each laser harmonic line in the longitudinal form factor spectrum

$$(\Delta\omega)_{\text{FWHM}} = 2\sqrt{2\ln 2}\frac{c/\sigma_z}{\sqrt{2}} = \frac{4\sqrt{2}\ln 2}{(\Delta t)_{\text{FWHM}}},$$
(4.20)

where Δt is the electron bunch length in unit of time. Then the relative bandwidth of the longitudinal form factor at the H-th laser harmonic can be expressed as

$$\left(\frac{\Delta\omega}{\omega}\right)_{\text{FWHM}} = \frac{2\sqrt{2}\ln 2}{\pi}\frac{1}{(c\Delta t)_{\text{FWHM}}/\lambda} = \frac{2\sqrt{2}\ln 2}{\pi}\frac{\lambda_L}{H(c\Delta t)_{\text{FWHM}}}. \tag{4.21}$$

Note that the coherent radiation pulse length is $\frac{1}{\sqrt{2}}$ of the electron bunch length due to the scaling of $P_{\text{coh}} \propto N_e^2$, and the above formula means that the coherent radiation is Fourier-transform limited. Note also that the absolute width $(\Delta\omega)_{\text{FWHM}}$ is independent of H, while the relative bandwidth $\left(\frac{\Delta\omega}{\omega}\right)_{\text{FWHM}} \propto \frac{1}{H}$.

4.2.2 Transverse From Factor

Now let us investigate the transverse form factor. Since the transverse form factor depends on the radiation process, there is not a universal formula involving only the beam distribution. Here for our interest, we focus on the case of undulator radiation. We use a planar undulator as an example. The formulation for a helical undulator is similar.

As well established in literature, the planar undulator radiation of a point charge in the H-th harmonic is [4]

$$\frac{d^2 W_H}{d\omega d\Omega}(\theta, \varphi, \omega)\bigg|_{\text{point}} = \frac{2e^2\gamma^2}{\pi\epsilon_0 c}G(\theta, \varphi)F(\epsilon),$$

$$F(\epsilon) = \left(\frac{\sin(\pi N_u\epsilon)}{\pi\epsilon}\right)^2, \quad \epsilon = \frac{\omega}{\omega_r(\theta)} - H = \frac{\omega}{2ck_u\gamma^2}\left(1 + K^2/2 + \gamma^2\theta^2\right) - H,$$

$$G(\theta, \varphi) = G_\sigma(\theta, \varphi) + G_\pi(\theta, \varphi),$$

$$G_\sigma(\theta, \varphi) = \left[\frac{H\left(\frac{K}{\sqrt{2}}\mathcal{D}_1 + \frac{\gamma\theta}{\sqrt{2}}\mathcal{D}_2\cos\varphi\right)}{1 + K^2/2 + \gamma^2\theta^2}\right]^2, \quad G_\pi(\theta, \varphi) = \frac{1}{2}\left(\frac{H\gamma\theta\mathcal{D}_2\sin\varphi}{1 + K^2/2 + \gamma^2\theta^2}\right)^2, \tag{4.22}$$

$$\mathcal{D}_1 = -\frac{1}{2}\sum_{m=-\infty}^{\infty}J_{H+2m-1}(H\alpha)\left[J_m(H\zeta) + J_{m-1}(H\zeta)\right],$$

$$\mathcal{D}_2 = \sum_{m=-\infty}^{\infty}J_{H+2m}(H\alpha)J_m(H\zeta),$$

$$\alpha = \frac{2K\gamma\theta\cos\varphi}{1 + K^2/2 + \gamma^2\theta^2}, \quad \zeta = \frac{K^2/4}{1 + K^2/2 + \gamma^2\theta^2},$$

in which e is the elementary charge, γ is the Lorentz factor, ϵ_0 is the permittivity of free space, $\omega_r(\theta)$ is the fundamental resonant angular frequency at the observation with a polar angle of θ, $k_u = \lambda_u/2\pi$ is the wavenumber of the undulator, $K = \frac{eB_0}{m_e c k_u} = 0.934 \cdot B_0[\text{T}] \cdot \lambda_u[\text{cm}]$ is the dimensionless undulator parameter, with B_0 being the peak magnetic flux density of the undulator and m_e being the mass of an electron, J_m means the m-th order Bessel function of the first kind.

Now we try to get some analytical results for the transverse form factor. Instead of a general discussion, here we consider only the simplest case of a transverse round Gaussian beam, i.e,

$$|b_\perp(\theta, \varphi, \omega)|^2 = \exp\left[-\left(\frac{\omega}{c}\sigma_\perp \sin\theta\right)^2\right], \tag{4.23}$$

where $\sigma_\perp$ is the RMS transverse size of the electron beam. As the radiation is dominantly in the forward direction in relativistic case, and $e^{-\left(\frac{\omega}{c}\sigma_\perp \sin\theta\right)^2}$ approaches zero with the increase of θ, therefore in Eq. (4.12), the upper limit of θ in the integration can be effectively replaced by infinity, and $\sin\theta$ can be replaced by θ in $e^{-\left(\frac{\omega}{c}\sigma_\perp \sin\theta\right)^2}$. Then

$$\int_0^{2\pi} d\varphi \int_0^\pi \sin\theta d\theta e^{-\left(\frac{\omega}{c}\sigma_\perp \sin\theta\right)^2} \frac{d^2 W_H}{d\omega d\Omega}(\theta, \varphi, \omega)\Big|_{\text{point}}$$

$$\approx \int_0^{2\pi} d\varphi \int_0^\infty \theta d\theta e^{-\left(\frac{\omega}{c}\sigma_\perp \theta\right)^2} \frac{d^2 W_H}{d\omega d\Omega}(\theta, \varphi, \omega)\Big|_{\text{point}} \tag{4.24}$$

$$\approx \frac{e^2}{\pi\epsilon_0 c}\int_0^{2\pi} d\varphi G(\theta_1, \varphi)\int_0^\infty d(\gamma\theta)^2 e^{-\left(\frac{\omega}{c}\sigma_\perp \theta\right)^2} F(\epsilon),$$

where $\theta_1 = \frac{1}{\gamma}\sqrt{(1 + K^2/2)\left(\frac{H\omega_0}{\omega} - 1\right)}$, $\omega_0 = \omega_r(\theta = 0) = \frac{2\gamma^2}{1+K^2/2}\omega_u$. Here we have made use of the fact that there is only one value of θ, i.e., θ_1, that contributes significantly to the integration over the solid angle Ω due to the sharpness of $F(\epsilon)$ when the undulator period number $N_u \gg 1$, as the spectral width of $F(\epsilon)$ is $1/N_u$.

The transverse form factor corresponding to the H-th harmonic can thus be defined as

$$FF_\perp(H, \omega) = \frac{\int_0^\infty d(\gamma\theta)^2 e^{-\left(\frac{\omega}{c}\sigma_\perp \theta\right)^2} \text{sinc}^2(N_u \pi \epsilon)}{\int_0^\infty d(\gamma\theta)^2 \text{sinc}^2(N_u \pi \epsilon)}. \tag{4.25}$$

The radiation spectrum of the H-th harmonic is then

$$\frac{dW_H}{d\omega}\Big|_{\text{beam}} = N_e^2 FF_\perp(H, \omega) FF_z(\omega)\frac{dW_H}{d\omega}\Big|_{\text{point}}, \tag{4.26}$$

and the total radiation spectrum of an electron beam is

$$\frac{dW}{d\omega}\Big|_{\text{beam}} = \sum_{H=1}^\infty \frac{dW_H}{d\omega}\Big|_{\text{beam}}. \tag{4.27}$$

Denote

$$\kappa_1 \equiv N_u \pi \left(\frac{\omega}{\omega_0} - H \right), \quad \kappa_2 \equiv N_u \pi \frac{\omega}{\omega_0} \frac{1}{1 + K^2/2}, \quad \kappa_3 \equiv \left(\frac{\omega \sigma_\perp}{c \gamma} \right)^2, \quad (4.28)$$

then the denominator in Eq. (4.25) is

$$\mathcal{D}(H, \omega) = \int_0^\infty dx \, \text{sinc}^2 \, (\kappa_1 + \kappa_2 x) = \frac{\frac{\pi}{2} - \text{Si}\,(2\kappa_1) + \frac{\sin^2(\kappa_1)}{\kappa_1}}{\kappa_2}, \quad (4.29)$$

where $\text{Si}(x) = \int_0^x \frac{\sin(t)}{t} dt$ is the sine integral, and the numerator in Eq. (4.25) is

$$\mathcal{N}(H, \omega) = \int_0^\infty dx \, e^{-\kappa_3 x} \text{sinc}^2 \, (\kappa_1 + \kappa_2 x)$$

$$= \frac{e^{\frac{\kappa_1 \kappa_3}{\kappa_2}}}{4\kappa_1 \kappa_2^2} \left\{ -4\kappa_2 \sin^2(\kappa_1) e^{-\frac{\kappa_1 \kappa_3}{\kappa_2}} - 2\kappa_1 \kappa_2 i \left[\text{Ei}\left(2\kappa_1 i - \frac{\kappa_1 \kappa_3}{\kappa_2} \right) - \text{Ei}\left(-2\kappa_1 i - \frac{\kappa_1 \kappa_3}{\kappa_2} \right) \right] \right.$$

$$\left. + \kappa_1 \kappa_3 \left[\text{Ei}\left(2\kappa_1 i - \frac{\kappa_1 \kappa_3}{\kappa_2} \right) - 2\text{Ei}\left(-\frac{\kappa_1 \kappa_3}{\kappa_2} \right) + \text{Ei}\left(-2\kappa_1 i - \frac{\kappa_1 \kappa_3}{\kappa_2} \right) \right] \right\}, \quad (4.30)$$

where $\text{Ei}(x) = \int_{-\infty}^x \frac{e^t}{t} dt$ is the exponential integral. The transverse form factor is then

$$FF_\perp(H, \omega) = \frac{\mathcal{N}(H, \omega)}{\mathcal{D}(H, \omega)}. \quad (4.31)$$

When $\omega = H\omega_0$, then $\kappa_1 = 0$, the transverse form factor has a simpler form,

$$FF_\perp(S) \equiv FF_\perp(H, H\omega_0) = \frac{\int_0^\infty dx \, e^{-\kappa_3 x} \text{sinc}^2 \, (\kappa_2 x)}{\int_0^\infty dx \, \text{sinc}^2 \, (\kappa_2 x)}$$

$$= \frac{2}{\pi} \left[\tan^{-1}\left(\frac{1}{2S} \right) + S \ln \left(\frac{(2S)^2}{(2S)^2 + 1} \right) \right], \quad (4.32)$$

where

$$S(\sigma_\perp, L_u, \omega) = \frac{\kappa_3}{4\kappa_2} = \frac{\sigma_\perp^2 k_u \frac{\omega}{c}}{2N_u \pi} = \frac{\sigma_\perp^2 \frac{\omega}{c}}{L_u} \quad (4.33)$$

is the diffraction parameter, with $L_u = N_u \lambda_u$ being the length of the undulator. This form factor $FF_\perp(S)$ is a universal function and has been obtained before in Ref. [5]. Here we have reproduced the result following the general definition of the transverse form factor. The variable S is a parameter used to classify the diffraction limit of the beam,

$$FF_\perp(S) = \begin{cases} 1, & S \ll 1, \text{ below diffraction limit}, \\ \frac{1}{2\pi S}, & S \gg 1, \text{ above diffraction limit}. \end{cases} \quad (4.34)$$

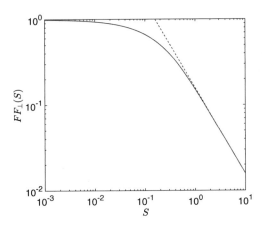

Fig. 4.3 The universal function $FF_\perp(S)$ and its asymptotic value above diffraction limit. The solid line comes from Eq. (4.32), the dashed line from the asymptotic relation above diffraction limit Eq. (4.34)

This function along with its asymptotic result above diffraction limit are shown in Fig. 4.3.

Note that the decrease of $FF_\perp(S)$ with the increase of $\sigma_\perp$ ($S \propto \sigma_\perp^2$) means that the coherent radiation at the frequency $\omega = H\omega_0$ becomes less when the transverse electron beam size becomes larger. This reflects the fact that for a given radiation frequency ω, there is a range of polar angle θ that can contribute. For $\omega = H\omega_0$, not only $\theta = 0$, but also θ very close to 0 contribute. With the increase of $\sigma_\perp$, the effective bunching factor $b(\theta, \varphi, \omega)$ at $\omega = H\omega_0$ drops for these non-zero θ due to the projected bunch lengthening, therefore the coherent radiation becomes less. Another way to appreciate the drop of $FF_\perp(S)$ with the increase of $\sigma_\perp$ is that there is a transverse coherence area whose radius is proportional to $\sqrt{L_u\lambda_0/H}$ with $\lambda_0 = 2\pi \frac{c}{\omega_0}$, and less particles can cohere with each other when the transverse size of the electron beam increases.

Note that our definition Eq. (4.12) and derivation of the transverse form factor Eq. (4.31) is more general than that given in Ref. [5], as it covers other frequencies in addition to a single frequency ω_0. Therefore, it contains more information than Eq. (4.32) as will be presented soon. However, Eq. (4.31) is still not simple enough for efficient analytical evaluation to provide physical insight. A further approximation is thus introduced,

$$
\begin{aligned}
FF_\perp(H, \omega) &= \frac{\int_0^\infty dx\, e^{-\kappa_3 x}\mathrm{sinc}^2\,(\kappa_1 + \kappa_2 x)}{\int_0^\infty dx\,\mathrm{sinc}^2\,(\kappa_1 + \kappa_2 x)} \\
&= e^{\frac{\kappa_1\kappa_3}{\kappa_2}} \frac{\int_{\kappa_1}^\infty dy\, e^{-\frac{\kappa_3}{\kappa_2}y}\mathrm{sinc}^2(y)}{\int_{\kappa_1}^\infty dy\,\mathrm{sinc}^2(y)} \\
&\approx e^{\frac{\kappa_1\kappa_3}{\kappa_2}} \frac{\int_0^\infty dy\, e^{-\frac{\kappa_3}{\kappa_2}y}\mathrm{sinc}^2(y)}{\int_0^\infty dy\,\mathrm{sinc}^2(y)} \\
&= e^{-4N_u\pi S\left(H-\frac{\omega}{\omega_0}\right)}FF_\perp(S).
\end{aligned}
\tag{4.35}
$$

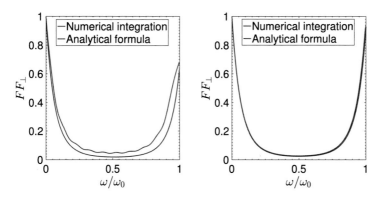

Fig. 4.4 The comparison of the transverse form factor, between that calculated from our simplified analytical formula Eq. (4.35) and that from the direct numerical integration of Eq. (4.12) for the case of $H = 1$, with $N_u = 10$ (left) and $N_u = 100$ (right), respectively. Other related parameters used in the calculation: $E_0 = 400$ MeV, $\lambda_0 = 13.5$ nm, $\lambda_u = 1$ cm, $K = 1.14$, $\sigma_\perp = 5$ μm

The condition of applying such simplification is $\frac{\kappa_3}{\kappa_2} (\omega = H\omega_0) \ll 1$ or $S (\omega = H\omega_0) \ll 1$, i.e., the beam is below diffraction limit for the on-axis radiation $\omega = H\omega_0$. Therefore, the conditions of applying Eq. (4.35) are

$$N_u \gg 1 \text{ and } \sigma_\perp \ll \frac{1}{\sqrt{H}} \sqrt{\frac{L_u \lambda_0}{2\pi}}. \qquad (4.36)$$

If the second condition is not satisfied, the more accurate result Eq. (4.31) should be referred.

As a benchmark of our derivation, here we conduct some calculations of the transverse form factor based on direct numerical integration of Eq. (4.12), and compare them with our simplified analytical formula Eq. (4.35). The parameters used are for the envisioned EUV SSMB to be presented in Sect. 4.6, and are given in the figure caption. As can be seen in Fig. 4.4, their agreement when $N_u = 100$ is remarkably well. Even in the case of $N_u = 10$, the agreement is still satisfactory. There are two reasons why the agreement is better in the case of a large N_u. The first is that in the derivation we have made use of the sharpness of $F(\epsilon)$, whose width is $1/N_u$. The second is that $S \propto \frac{1}{N_u}$ with a given transverse beam size and undulator period length, and our simplified analytical formula Eq. (4.35) applies when $S (\omega = H\omega_0) \ll 1$.

To appreciate the implication of the generalized transverse form factor further, an example flat contour plot of the transverse form factor as a function of the radiation frequency ω and transverse electron beam size $\sigma_\perp$ is shown in Fig. 4.5. As can be seen, a large transverse electron beam size will suppress the off-axis red-shifted coherent radiation due to the projected bunch lengthening from the transverse size, thus the effective bunching factor degradation, when observed off-axis. Therefore, a large transverse electron beam size will make the coherent radiation more collimated in the forward direction, and more narrowbanded around the harmonic lines. But note

Fig. 4.5 Flat contour plot of the transverse form factor $FF_\perp(H, \omega)$ for $H = 1$, as a function of the radiation frequency ω and transverse electron beam size $\sigma_\perp$, calculated using our simplified analytical formula Eq. (4.35). Parameters used in the calculation: $E_0 = 400$ MeV, $\lambda_0 = 13.5$ nm, $\lambda_u = 1$ cm, $K = 1.14$, $N_u = 2 \times 79$

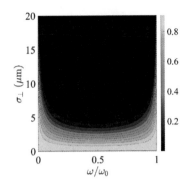

that not only the red-shifted radiation is suppressed, the radiation strength of each harmonic line $\omega = H\omega_0$ actually also decreases with the increase of the transverse electron beam size, the reason of which we have just explained.

Now we evaluate the bandwidth and opening angle of the radiation at different harmonics due to the transverse form factor. In particular, we are interested in the case where the off-axis red-shifted radiation is significantly suppressed by the transverse size of the electron beam, which requires that

$$N_u \pi S(\omega = H\omega_0) \gg 1, \tag{4.37}$$

i.e., $\sigma_\perp \gg \sqrt{\frac{H}{2}} \frac{\sqrt{\lambda_u \lambda_0}}{2\pi}$. Note that to apply Eq. (4.35), we still need the conditions in Eq. (4.36). For example, to apply the analytical estimation for the example EUV SSMB calculation to be presented in Sect. 4.6, in which $\lambda_u = 1$ cm, $\lambda_0 = 13.5$ nm and $N_u = 2 \times 79$, we need 1.3 μm $\ll \sigma_\perp \ll 58$ μm. The typical transverse electron beam size in an EUV SSMB ring is in this range.

With these conditions satisfied, the value of the exponential factor in Eq. (4.35) is more sensitive to the change of ω, compared to the universal function $FF_\perp(S)$. Therefore, here we consider only the exponential term when ω is close to $H\omega_0$. We want to know the ω at which the exponential term gives $e^{-4N_u \pi S\left(H - \frac{\omega}{\omega_0}\right)} = e^{-1}$.

Putting in the definition of $S = \frac{\sigma_\perp^2 k_u \frac{\omega}{c}}{2N_u \pi}$, we have $\omega_{e^{-1}} = \frac{1 + \sqrt{1 - \frac{2}{H^2 \sigma_\perp^2 k_u k_0}}}{2} H\omega_0$. Then

$$\Delta\omega_{e^{-1}}\bigg|_\perp = H\omega_0 - \omega_{e^{-1}} = \frac{1 - \sqrt{1 - \frac{2}{H^2 \sigma_\perp^2 k_u k_0}}}{2} H\omega_0. \tag{4.38}$$

As $\sigma_\perp \gg \sqrt{\frac{H}{2}} \frac{\sqrt{\lambda_u \lambda_0}}{2\pi}$, therefore $\frac{2}{H^2 \sigma_\perp^2 k_u k_0} \ll 1$, we have

$$\frac{\Delta\omega_{e^{-1}}}{H\omega_0}\bigg|_\perp \approx \frac{1}{2H^2 \sigma_\perp^2 k_u k_0}. \tag{4.39}$$

Correspondingly, the opening angle of the coherent radiation due to the transverse form factor is

$$\frac{\gamma^2 \theta_{e-1}^2}{1 + K^2/2} \approx \left. \frac{\Delta \omega_{e-1}}{H \omega_0} \right|_{\perp} \Rightarrow \left. \theta_{e-1} \right|_{\perp} \approx \frac{\sqrt{2 + K^2}}{2 H \gamma \sigma_{\perp} \sqrt{k_u k_0}}. \tag{4.40}$$

It is interesting to note that

$$\left. \frac{\Delta \omega_{e-1}}{H \omega_0} \right|_{\perp} \propto \frac{1}{H^2}. \tag{4.41}$$

As a comparison, the relative bandwidth at the harmonics due to the longitudinal form factor is

$$\left. \frac{\Delta \omega}{H \omega_0} \right|_z \propto \frac{1}{H}. \tag{4.42}$$

Note also that $\left. \frac{\Delta \omega_{e-1}}{H \omega_0} \right|_{\perp}$ and $\left. \theta_{e-1} \right|_{\perp}$ are independent of N_u, although the approximations adopted in the derivation actually involve conditions related to N_u.

4.3 Radiation Power and Spectral Flux

In many cases, the microbunching is formed based on an electron bunch much longer than the radiation wavelength, for example in an FEL or CHG. In these cases, the linewidth of the longitudinal form factor at the harmonics are usually even narrower than that given by the transverse form factor. This also means that the coherent radiation of a long continuous electron bunch-based microbunching will be dominantly in the forward direction, as the bunching factor of the off-axis red-shifted frequency is suppressed very fast compared to the on-axis resonant ones. For a more practical application, here we derive the coherent radiation power and spectral flux at the undulator radiation harmonics in these cases. As we will see, the results can be viewed as useful references for SSMB radiation.

We assume that the long electron bunch, before microbunching, is Gaussain. The transverse form factors around the harmonics do not change much, i.e., we assume $e^{-4N_u \pi S \left(H - \frac{\omega}{\omega_0} \right)} \approx 1$ when ω is close to $H \omega_0$. Therefore, we only need to take into account the Gaussian shape of the longitudinal form factors at the harmonics. The RMS bandwidth of the longitudinal form factor for a Gaussian bunch with a length of σ_z is $\Delta \omega|_z = \frac{c}{\sigma_z/\sqrt{2}}$. Therefore, the coherent radiation energy at the H-th harmonic is

$$W_H = \left[N_e^2 FF_\perp(H,\omega) FF_z(\omega) \frac{dW_H}{d\omega}\bigg|_{\text{point}} \right] (\omega = H\omega_0) \times \int_{-\infty}^{\infty} \exp\left(-\frac{(\omega - H\omega_0)^2}{(c/\sigma_z)^2} \right) d\omega$$

$$= N_e^2 FF_\perp(S)|b_{z,H}|^2 \frac{2e^2}{\epsilon_0 c} G(\theta = 0) \frac{1 + K^2/2}{H} \frac{N_u}{2} \times \sqrt{\pi} c/\sigma_z.$$

$$(4.43)$$

For a planar undulator, the σ-mode radiation dominates and from Eq. (4.22) we have

$$G_\sigma(\theta = 0) = \left[\frac{HK/\sqrt{2}}{2(1 + K^2/2)} \right]^2 [JJ]_H^2, \qquad (4.44)$$

in which the denotation $[JJ]_H^2 = \left[J_{\frac{H-1}{2}}(H\chi) - J_{\frac{H+1}{2}}(H\chi) \right]^2$, with $\chi = \frac{K^2}{4+2K^2}$, is used. Note however, the above expression is meaningful only for an odd H, as the on-axis even harmonic radiation is rather weak. The peak power of the odd-H-th harmonic coherent radiation is then

$$P_{H,\text{peak}} = \frac{W_H}{\sqrt{2\pi} \frac{\sigma_z/c}{\sqrt{2}}} = \frac{\pi}{\epsilon_0 c} N_u H\chi [JJ]_H^2 FF_\perp(S)|b_{z,H}|^2 I_P^2, \qquad (4.45)$$

where $I_P = \frac{N_e e}{\sqrt{2\pi}\sigma_z/c}$ is the peak current of the Gaussian bunch before microbunching. For a more practical application of the derived formula, we put in the numerical value of the constants, and arrive at

$$P_{H,\text{peak}}[\text{kW}] = 1.183 N_u H\chi [JJ]_H^2 FF_\perp(S)|b_{z,H}|^2 I_P^2[\text{A}]. \qquad (4.46)$$

Note that the above formula applies when the radiation slippage length $N_u\lambda_0$ is smaller than the bunch length σ_z. If not, the above formula will overestimate the coherent radiation peak power, as the RMS radiation pulse length is then longer than $\frac{c/\sigma_z}{\sqrt{2}}$. Note also that given the same bunch charge and form factors, $P_{H,\text{peak}} \propto I_P^2 \propto \frac{1}{\sigma_z^2}$ while $W_H \propto \frac{1}{\sigma_z}$. The reason a shorter bunch radiates more total energy is because that more particles are within the coherence length.

At a first glance of Eq. (4.45), the coherent radiation power P_{coh} seems to be proportional to N_u, while an intuitive picture of the longitudinal coherence length $l_{\text{coh}} \propto N_u$ says that the scaling should be $P_{\text{coh}} \propto N_u^2$, as the electron number within the coherence length is proportional to N_u. This is actually because that $FF_\perp(S)$ is also a function of N_u. It is interesting to note that

$$P_{\text{coh}} = \begin{cases} \propto N_u, & \text{below diffraction limit,} \\ \propto N_u^2, & \text{above diffraction limit,} \end{cases} \qquad (4.47)$$

which can be obtained from the asymptotic expressions of $FF_\perp(S)$ as shown in Eq. (4.34). So for a given transverse beam size, $P_{\text{coh}} \propto N_u^2$ at first when N_u is small. When N_u is large enough such that the electron beam is below diffraction limit, then $P_{\text{coh}} \propto N_u$. Physically this is because with the increase of N_u, the diffraction of the radiation will prevent the radiation from one particle so effectively affect the particles far in front of it, as the on-axis field from this particle becomes weaker with the increase of the radiation slippage length.

Our derivation of the coherent radiation power above is for a Gaussian bunch-based microbunching. For a coasting or DC beam, we just need to replace I_P in Eq. (4.45) by the average current I_A, and the peak power is then the average power. For a helical undulator, we need to replace $K_{\text{planar}}/\sqrt{2}$ with K_{helical}, and $[JJ]_1^2$ with 1, in the evaluation of the radiation power at fundamental frequency.

We remind the readers that Eq. (4.45) is for the case of a long continuous bunch-based microbunching, for example in FELs and CHG. In some of the SSMB scenarios, the microbunches are cleanly separated from each other according to the modulation laser wavelength as will be shown in Fig. 4.9, and usually the radiation wavelength is at a high harmonic of the modulation laser. Therefore, there actually could be significant red-shifted radiation generated in SSMB as we will see in the example calculation in Sect. 4.6. If we put the average current of SSMB in Eq. (4.45), what it evaluates is the radiation power whose frequency content is close to the on-axis harmonic and will underestimate the overall radiation power.

After investigating the radiation power, let us now have a look at the spectral flux, which is the number of photons per unit time in a given small bandwidth. The spectral flux of coherent radiation at the odd-H-th harmonic can be calculated according to $\frac{dW_H}{d\omega}$ as

$$
\mathcal{F}(\omega = H\omega_0) = \left[N_e^2 FF_\perp(H, \omega) FF_z(\omega) \frac{dW_H}{d\omega}\bigg|_{\text{point}} \times \frac{\Delta\omega}{\hbar\omega} \right] (\omega = H\omega_0)
$$

$$
= \frac{1}{1000} \frac{e^2}{2\epsilon_0 c\hbar} N_u H \chi [JJ]_H^2 FF_\perp(S) |b_{z,H}|^2 N_e^2 \text{ (phs/pass/0.1\%b.w.)},
$$
(4.48)

where $\hbar$ is the reduced Planck's constant. Again we put in the numerical value of the constants, and arrive at

$$
\mathcal{F}(\omega = H\omega_0) = 4.573 \times 10^{-5} N_u H \chi [JJ]_H^2 FF_\perp(S) |b_{z,H}|^2 N_e^2 \text{ (phs/pass/0.1\%b.w.)}.
$$
(4.49)

Note that the above spectral flux is for a single pass of the microbunched electron beam through the radiator undulator. For the evaluation of the average spectral flux in an SSMB storage ring, we need to multiply it with the number of microbunches passing a fixed location in one second, namely $F \frac{\bar{v}_z}{\lambda_L}$, with F being the filling factor of microbunches in the ring, $\bar{v}_z$ being the average longitudinal speed of electron wiggling in the modulator undulator, and λ_L being the modulation laser wavelength. We remind the readers that the above statement means we do not account for the radiation overlapping between different microbunches if the radiation slippage length is larger than $\frac{c}{\bar{v}_z}\lambda_L$. If there is such radiation overlapping, the flux will be boosted further since the electrons in neighboring microbunches can now cohere with each other. To give the readers a more concrete feeling about the high spectral flux in SSMB, we just need to multiply the spectral flux of the usual incoherent undulator radiation with a factor of $N_e F F_\perp(S)|b_{z,H}|^2$, with N_e being the number of electrons per microbunch. For example, in the envisioned EUV SSMB to be presented in Sect. 4.6, $N_e = 2.2 \times 10^4$, and $F F_\perp(S)|b_{z,H}|^2$ can be as large as 0.1. Therefore, the EUV spectral flux in an SSMB storage ring can thus be three orders of magnitude higher than that in a conventional synchrotron source.

4.4 Impact of Electron Beam Divergence and Energy Spread

Now we take into account the impact of beam divergence and energy spread on the coherent radiation. With an aim to obtain some efficient evaluation, here we simplify the analysis by considering only the impact of particle's x', y' and δ on the arrival time of the radiation, not on the radiation pattern. This approximation is valid when the beam divergence and energy spread are small enough, such that $\sigma_{x',y'} < \frac{1}{\gamma}$ and $\sigma_\delta < \frac{1}{N_u}$. Basically we want to get a formula of the effective transverse and longitudinal form factors considering the beam size evolution during radiation.

As an example, here we assume that the beam is a 6D Gaussian one, and round in the transverse dimension. Further we assume the beam reaches its minimal in all three dimensions at the radiator undulator center, which is desired to get high-power radiation, then the effective transverse and longitudinal form factors are

$$FF_\perp(\omega) = \frac{1}{L_u} \int_{-\frac{L_u}{2}}^{\frac{L_u}{2}} FF_\perp \left(\frac{(\sigma_\perp^2 + (\sigma_{\theta_\perp} s)^2) \frac{\omega}{c}}{L_u} \right) ds,$$

$$FF_z(\omega) = \frac{1}{L_u} \int_{-\frac{L_u}{2}}^{\frac{L_u}{2}} e^{-\left(\frac{\omega}{c}\right)^2 \left[\sigma_z^2 + \left(\sigma_\delta \frac{s}{L_u 2 N_u \lambda_0}\right)^2\right]} ds \qquad (4.50)$$

$$= e^{-\left(\frac{\omega}{c}\right)^2 \sigma_z^2} \frac{\sqrt{\pi}}{2} \frac{\mathrm{erf}\left(\frac{\omega}{c}\sigma_\delta N_u \lambda_0\right)}{\frac{\omega}{c}\sigma_\delta N_u \lambda_0},$$

where $\sigma_\perp$, $\sigma_{\theta_\perp}$, σ_z and σ_δ are the transverse beam size, divergence, bunch length and energy spread at the undulator center, with $\sigma_\perp \sigma_{\theta_\perp} = \epsilon_\perp$ and $\sigma_z \sigma_\delta = \epsilon_z$, and $\mathrm{erf}(x) = \frac{2}{\sqrt{\pi}} \int_0^x e^{-t^2} dt$ is the error function. Note that $\frac{\sqrt{\pi}}{2} \frac{\mathrm{erf}(x)}{x} < 1$ for $x \neq 0$.

Figure 4.6 is an example plot of the effective form factors based on the above formulas, and the comparison with the case for a 3D rigid beam. It can be seen that given a transverse and longitudinal emittance, there is an optimal transverse beam size and bunch length at the radiator center considering the impact of beam divergence and energy spread. This is expected, since the beam size or bunch length of an over-focused beam grows very fast.

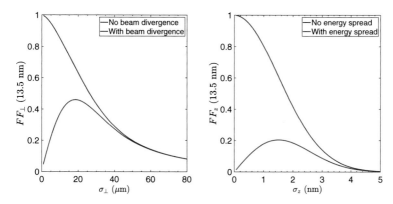

Fig. 4.6 An example plot to show the optimization of beam sizes in the middle of the radiator, considering the beam divergence and energy spread. Parameters used for calculation: $\epsilon_{\perp} = 1$ nm, $\epsilon_z = 4$ pm, $\lambda_0 = 13.5$ nm, $N_u = 2 \times 79$, $L_u = 1.58$ m

4.5 Statistical Property of Radiation

In the previous sections, we have ignored the quantum discrete nature of radiation. Besides, we have derived the coherent radiation property using a smooth distributed charge, i.e., we have treated the charge as a continuum fluid. The number of photons radiated from a charged particle beam actually fluctuates from turn to turn or bunch to bunch if the quantum nature of radiation and the pointlike nature of electrons are taken into account. The first mechanism exists even if there is only one electron, and the second mechanism is related to the interference of fields radiated by different electrons [6]. Using the classical language, the second fluctuation mechanism is from the fluctuation of the bunching factor or form factor of electron beam.

There have been studies on the statistical property of the radiation in FELs [7] and also the storage ring-based synchrotron radiation sources [6, 8]. Rich information about the electron beam is embedded in the radiation fluctuations, or more generally the statistical property of the radiation. For example, the turn-by-turn fluctuation of the incoherent undulator radiation can be used to measure the transverse emittance of electron beam [8]. The previous treatment, however, usually cares about the cases where the bunch length is much longer than the radiation wavelength, i.e., the radiation is temporally incoherent (in SASE FEL, incoherent at the beginning). In SSMB, the bunch length is comparable or shorter than the desired radiation wavelength, and the dominant radiation is temporally coherent. Although numerical calculation is doable following the general theoretical formulation, an analytical formula for the fluctuation in this temporally coherent radiation dominant regime is of value for a better understanding of the physics and investigation of its potential applications.

4.5.1 Pointlike Nature of Electron

Here for SSMB we consider first the second mechanism of fluctuation, i.e., the radiation power fluctuation arising from the pointlike nature of the radiating electron. In this section, to simplify writing, we use the vector notation

$$
\begin{aligned}
\mathbf{k} &= \frac{\omega}{c} \left(\sin\theta \cos\varphi, \sin\theta \sin\varphi, 1 \right), \\
\mathbf{r} &= (x, y, z).
\end{aligned}
\tag{4.51}
$$

Then the bunching factor with the pointlike nature of electrons taken into account is

$$
b(\mathbf{k}) = b(\theta, \varphi, \omega) = \frac{1}{N_e} \sum_{n=1}^{N_e} e^{-i\mathbf{k}\cdot\mathbf{r_n}}.
\tag{4.52}
$$

First we want to evaluate the coherent radiation power fluctuation at a specific frequency and observation angle. As the radiation power is proportional to $N_e^2 |b(\mathbf{k})|^2$, therefore we need to know the fluctuation of $|b(\mathbf{k})|^2$. Since

$$
|b(\mathbf{k})|^2 = \frac{1}{N_e^2} \sum_{n=1}^{N_e} \sum_{m=1}^{N_e} e^{-i\mathbf{k}\cdot(\mathbf{r}_n - \mathbf{r}_m)} = \frac{1}{N_e^2} \left[N_e + \sum_{m\neq n} e^{-i\mathbf{k}\cdot(\mathbf{r}_n - \mathbf{r}_m)} \right],
\tag{4.53}
$$

we have

$$
\langle |b(\mathbf{k})|^2 \rangle = \frac{1}{N_e} + \left(1 - \frac{1}{N_e} \right) |\bar{b}(\mathbf{k})|^2,
\tag{4.54}
$$

with

$$
\bar{b}(\mathbf{k}) = \bar{b}(\theta, \varphi, \omega) = \int \rho(\mathbf{r}) e^{-i\mathbf{k}\cdot\mathbf{r}} d\mathbf{r}
\tag{4.55}
$$

being the bunching factor we calculated before using a continuum fluid charge distribution in Eq. (4.3). As can be seen from Eq. (4.54), when $N_e = 1$, which corresponds to the case of a single point charge, then $\langle |b(\mathbf{k})|^2 \rangle = 1$. When $N_e \gg 1$ and $N_e |\bar{b}(\mathbf{k})|^2 \ll 1$, which corresponds to the case of incoherent radiation dominance, then $\langle |b(\mathbf{k})|^2 \rangle = \frac{1}{N_e}$. When $N_e \gg 1$ and $N_e |\bar{b}(\mathbf{k})|^2 \gg 1$, which corresponds to the case of coherent radiation dominance, then $\langle |b(\mathbf{k})|^2 \rangle = |\bar{b}(\mathbf{k})|^2$. These results are as expected.

The calculation of $\langle |b(\mathbf{k}_1)|^2 |b(\mathbf{k}_2)|^2 \rangle$ is more involved. More specifically,

$$
|b(\mathbf{k}_1)|^2 |b(\mathbf{k}_2)|^2 = \frac{1}{N_e^4} \sum_{n=1}^{N_e} \sum_{m=1}^{N_e} \sum_{p=1}^{N_e} \sum_{q=1}^{N_e} e^{-i\mathbf{k}_1\cdot(\mathbf{r}_n - \mathbf{r}_m) - i\mathbf{k}_2\cdot(\mathbf{r}_p - \mathbf{r}_q)}.
\tag{4.56}
$$

Table 4.1 The N_e^4 terms in the quadruple sum of Eq. (4.56) can be placed in 15 different classes, as shown in Ref. [9]

Item number	Index relations	Number of terms
(1)	$n = m = p = q$	N_e
(2)	$n = m, p = q, n \neq p$	$N_e(N_e - 1)$
(3)	$n = m, p \neq q \neq n$	$N_e(N_e - 1)(N_e - 2)$
(4)	$n = p, m = q, n \neq m$	$N_e(N_e - 1)$
(5)	$n = p, m \neq q \neq n$	$N_e(N_e - 1)(N_e - 2)$
(6)	$n = q, m = p, n \neq m$	$N_e(N_e - 1)$
(7)	$n = q, m \neq p \neq n$	$N_e(N_e - 1)(N_e - 2)$
(8)	$n = m = p, n \neq q$	$N_e(N_e - 1)$
(9)	$n = m = q, n \neq p$	$N_e(N_e - 1)$
(10)	$n = p = q, n \neq m$	$N_e(N_e - 1)$
(11)	$p = q = m, n \neq m$	$N_e(N_e - 1)$
(12)	$n \neq m \neq p \neq q$	$N_e(N_e - 1)(N_e - 2)(N_e - 3)$
(13)	$p = q, n \neq m \neq p$	$N_e(N_e - 1)(N_e - 2)$
(14)	$m = q, n \neq m \neq p$	$N_e(N_e - 1)(N_e - 2)$
(15)	$m = p, n \neq m \neq q$	$N_e(N_e - 1)(N_e - 2)$

The N_e^4 terms in this summation can be placed in 15 different cases, as shown in Table 4.1. Corresponding to the 15 cases, we have

$$
\begin{aligned}
\left\langle |b(\mathbf{k_1})|^2 |b(\mathbf{k_2})|^2 \right\rangle = \frac{1}{N_e^4} \Big[&N_e \overline{b}(\mathbf{k_1} - \mathbf{k_1} + \mathbf{k_2} - \mathbf{k_2}) \\
&+ N_e(N_e - 1)\overline{b}(\mathbf{k_1} - \mathbf{k_1})\overline{b}(\mathbf{k_2} - \mathbf{k_2}) \\
&+ N_e(N_e - 1)(N_e - 2)\overline{b}(\mathbf{k_1} - \mathbf{k_1})\overline{b}(\mathbf{k_2})\overline{b}(-\mathbf{k_2}) \\
&+ N_e(N_e - 1)\overline{b}(\mathbf{k_1} + \mathbf{k_2})\overline{b}(-\mathbf{k_1} - \mathbf{k_2}) \\
&+ N_e(N_e - 1)(N_e - 2)\overline{b}(\mathbf{k_1} + \mathbf{k_2})\overline{b}(-\mathbf{k_1})\overline{b}(-\mathbf{k_2}) \\
&+ N_e(N_e - 1)\overline{b}(\mathbf{k_1} - \mathbf{k_2})\overline{b}(-\mathbf{k_1} + \mathbf{k_2}) \\
&+ N_e(N_e - 1)(N_e - 2)\overline{b}(\mathbf{k_1} - \mathbf{k_2})\overline{b}(-\mathbf{k_1})\overline{b}(\mathbf{k_2}) \\
&+ N_e(N_e - 1)\overline{b}(\mathbf{k_1} - \mathbf{k_1} + \mathbf{k_2})\overline{b}(-\mathbf{k_2}) \\
&+ N_e(N_e - 1)\overline{b}(\mathbf{k_1} - \mathbf{k_1} - \mathbf{k_2})\overline{b}(\mathbf{k_2}) \\
&+ N_e(N_e - 1)\overline{b}(\mathbf{k_1} + \mathbf{k_2} - \mathbf{k_2})\overline{b}(-\mathbf{k_1}) \\
&+ N_e(N_e - 1)\overline{b}(-\mathbf{k_1} + \mathbf{k_2} - \mathbf{k_2})\overline{b}(\mathbf{k_1}) \\
&+ N_e(N_e - 1)(N_e - 2)(N_e - 3)\overline{b}(\mathbf{k_1})\overline{b}(-\mathbf{k_1})\overline{b}(\mathbf{k_2})\overline{b}(-\mathbf{k_2}) \\
&+ N_e(N_e - 1)(N_e - 2)\overline{b}(\mathbf{k_1})\overline{b}(-\mathbf{k_1})\overline{b}(\mathbf{k_2} - \mathbf{k_2}) \\
&+ N_e(N_e - 1)(N_e - 2)\overline{b}(\mathbf{k_1})\overline{b}(\mathbf{k_2})\overline{b}(-\mathbf{k_1} - \mathbf{k_2}) \\
&+ N_e(N_e - 1)(N_e - 2)\overline{b}(\mathbf{k_1})\overline{b}(-\mathbf{k_2})\overline{b}(-\mathbf{k_1} + \mathbf{k_2}) \Big].
\end{aligned}
\tag{4.57}
$$

The above result can be re-organized as

$$\langle |b(\mathbf{k_1})|^2 |b(\mathbf{k_2})|^2 \rangle = \frac{1}{N_e^4} \Big[N_e^2 + N_e^2(N_e - 1)\left(|\bar{b}(\mathbf{k_1})|^2 + |\bar{b}(\mathbf{k_2})|^2\right)$$
$$+ N_e(N_e - 1)(N_e - 2)\left(\bar{b}(\mathbf{k_1} + \mathbf{k_2})\bar{b}(-\mathbf{k_1})\bar{b}(-\mathbf{k_2}) + \text{c.c.}\right)$$
$$+ N_e(N_e - 1)(N_e - 2)\left(\bar{b}(\mathbf{k_1} - \mathbf{k_2})\bar{b}(-\mathbf{k_1})\bar{b}(\mathbf{k_2}) + \text{c.c.}\right)$$
$$+ N_e(N_e - 1)\left(|\bar{b}(\mathbf{k_1} + \mathbf{k_2})|^2 + |\bar{b}(\mathbf{k_1} - \mathbf{k_2})|^2\right)$$
$$+ N_e(N_e - 1)(N_e - 2)(N_e - 3)|\bar{b}(\mathbf{k_1})|^2|\bar{b}(\mathbf{k_2})|^2 \Big], \tag{4.58}$$

in which c.c. means complex conjugate.

If $\mathbf{k_1} = \mathbf{k_2} = \mathbf{k}$, then

$$\langle |b(\mathbf{k})|^4 \rangle - \langle |b(\mathbf{k})|^2 \rangle^2 = \frac{1}{N_e^4} \Big[N_e(N_e - 1) + 2N_e(N_e - 1)(N_e - 2)|\bar{b}(\mathbf{k})|^2$$
$$+ 2N_e(N_e - 1)(N_e - 2)\text{Re}\left[\bar{b}(2\mathbf{k})\bar{b}^2(-\mathbf{k})\right] \tag{4.59}$$
$$+ N_e(N_e - 1)|\bar{b}(2\mathbf{k})|^2$$
$$- 2N_e(N_e - 1)(2N_e - 3)|\bar{b}(\mathbf{k})|^4 \Big],$$

where Re[] means taking the real part of a complex number.

When $N_e \gg 1$ and $N_e|b(\mathbf{k})|^2 \ll 1$, which is the case for incoherent radiation dominance, we have $\langle |b(\mathbf{k})|^2 \rangle = \frac{1}{N_e}$ and

$$\frac{\text{Var}\left[|b(\mathbf{k})|^2\right]}{\langle |b(\mathbf{k})|^2 \rangle^2} = 1 + O\left(\frac{1}{N_e}\right), \tag{4.60}$$

where Var[] means the variance of, and $O(x^n)$ means terms of order x^n and higher. Therefore, the relative fluctuation of incoherent radiation is relatively large. This is also the reason why SASE-FEL radiation has a large shot-to-shot power fluctuation.

When $N_e \gg 1$ and $N_e|b(\mathbf{k})|^2 \gg 1$, which corresponds to the case of coherent radiation dominance like that in SSMB, we have

$$\frac{\text{Var}\left[|b(\mathbf{k})|^2\right]}{\langle |b(\mathbf{k})|^2 \rangle^2} = \frac{2}{N_e}\left(\frac{|\bar{b}(\mathbf{k})|^2 + \text{Re}\left[\bar{b}(2\mathbf{k})\bar{b}^2(-\mathbf{k})\right]}{|\bar{b}(\mathbf{k})|^4} - 2\right) + O\left(\frac{1}{N_e^2}\right). \tag{4.61}$$

The above equation is the main result of our analysis of bunching factor fluctuation for the regime of coherent radiation dominance, and to our knowledge is new. The formula can be used to evaluate coherent radiation power fluctuation at a specific frequency and observation angle. If the transverse electron beam size is zero, or if we observe on-axis, then we can just replace $b(\mathbf{k})$ with $b_z(\omega)$ in the above formula. We remind the readers that $b(\mathbf{k})$ in general is a complex number.

Now we conduct some numerical simulations to confirm our analysis of coherent radiation fluctuation. As can be seen from Fig. 4.7, which correspond to the cases of

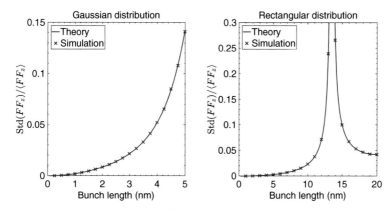

Fig. 4.7 Fluctuation of the longitudinal form factor at 13.5 nm v.s. bunch length with $N_e = 2.2 \times 10^4$. The bunch distribution is assumed to be Gaussian in the left and rectangular in the right, and the theoretical fluctuation is calculated according to Eq. (4.61), omitting the term $O\left(\frac{1}{N_e^2}\right)$. For each parameters choice, 1×10^4 simulations have been conducted to obtain the fluctuation

a Gaussian and a rectangular distributed bunch, respectively, the simulation results agree well with our theoretical prediction.

After investigating the expectation and variance of $|b(\mathbf{k})|^2$, one may be curious about its more detailed distribution. It can be shown that when $N_e|b(\mathbf{k})|^2 \gg 1$, the distribution of $|b(\mathbf{k})|^2$ tends asymptotically toward Gaussian.

As explained before, for a fixed frequency ω, there is a range of polar angle θ which can contribute. To evaluate the overall radiation power fluctuation at a specific frequency ω, we then need to know the fluctuation of the form factor $FF(\omega)$ which involves calculation depending on the specific radiation process. For our interested undulator radiation, it seems not easy to get a concise closed-form analytical formula to evaluate the total radiation power fluctuation when the beam has a finite transverse beam size. So here we refer to numerical calculation to give the readers a more concrete feeling about the impact of transverse size on coherent radiation power fluctuation.

For simplicity, we assume that the bunch length is zero and focus on the fluctuation of the transverse form factor. As can be seen from the simulation result in Fig. 4.8, the larger the transverse beam size, the larger relative fluctuation of the transverse form factor. We also observe that in a typical parameters set of the envisioned EUV SSMB, the fluctuation of 13.5 nm radiation power due to the finite transverse size is small. For example if $\sigma_\perp = 16\,\mu$m, then the relative fluctuation of the transverse form factor as shown in Fig. 4.8 is 0.14%. While the relative fluctuation of the longitudinal form factor at 13.5 nm when $\sigma_z = 3$ nm according to Eq. (4.61) is about 2%. Assuming that the beam is transverse-longitudinal decoupled, then

$$\frac{\text{Var}[FF(\omega)]}{\langle FF(\omega)\rangle^2} = \frac{\text{Var}[FF_\perp]}{\langle FF_\perp\rangle^2} + \frac{\text{Var}[FF_z]}{\langle FF_z\rangle^2} + \frac{\text{Var}[FF_\perp]\text{Var}[FF_z]}{\langle FF_\perp\rangle^2\langle FF_z\rangle^2}. \tag{4.62}$$

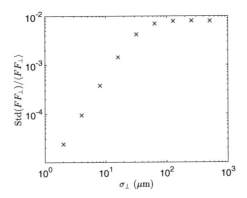

Fig. 4.8 Fluctuation of the transverse form factor at 13.5 nm v.s. transverse beam size with $N_e = 2.2 \times 10^4$. The bunch is assumed to have zero length and is round Gaussian in the transverse plane. For each parameters choice, 1×10^3 simulations have been conducted to obtain the fluctuation. Parameters used for the calculation: $E_0 = 400$ MeV, $\lambda_u = 1$ cm, $K = 1.14$, $N_u = 2 \times 79$

Therefore, for the envisioned EUV SSMB, the fluctuation of the longitudinal form factor dominates.

After investigating the power fluctuation at a specific frequency ω, now we look into the radiation energy fluctuation gathered within a finite frequency bandwidth and a finite angle acceptance. We use a filter function of $FT(\theta, \varphi, \omega)$ to account for the general case of frequency filter, angle acceptance, and detector efficiency. The expectation of the gathered photon energy and photon energy squared are

$$\langle W \rangle = N_e^2 \int_0^\pi \sin\theta d\theta \int_0^{2\pi} d\varphi \int_0^\infty d\omega FT(\theta, \varphi, \omega) \frac{d^2 W}{d\omega d\Omega}(\theta, \varphi, \omega)\bigg|_{\text{point}} \langle |b(\theta, \varphi, \omega)|^2 \rangle \tag{4.63}$$

and

$$\langle W^2 \rangle = N_e^4 \int_0^\pi \sin\theta d\theta \int_0^{2\pi} d\varphi \int_0^\infty d\omega \int_0^\pi \sin\theta' d\theta' \int_0^{2\pi} d\varphi' \int_0^\infty d\omega'$$

$$FT(\theta, \varphi, \omega) FT(\theta', \varphi', \omega') \frac{d^2 W}{d\omega d\Omega}(\theta, \varphi, \omega)\bigg|_{\text{point}} \frac{d^2 W}{d\omega d\Omega}(\theta', \varphi', \omega')\bigg|_{\text{point}}$$

$$\langle |b(\theta, \varphi, \omega)|^2 \rangle \langle |b(\theta', \varphi', \omega')|^2 \rangle g_2(\theta, \theta', \varphi, \varphi', \omega, \omega'), \tag{4.64}$$

where

$$g_2(\theta, \theta', \varphi, \varphi', \omega, \omega') = \frac{\langle |b(\theta, \varphi, \omega)|^2 |b(\theta', \varphi', \omega')|^2 \rangle}{\langle |b(\theta, \varphi, \omega)|^2 \rangle \langle |b(\theta', \varphi', \omega')|^2 \rangle}, \tag{4.65}$$

whose calculation can follow similar approach of calculating $\langle |b(\mathbf{k_1})|^2 |b(\mathbf{k_2})|^2 \rangle$ in Eq. (4.57). And the relative fluctuation of the gathered photon energy is

$$\sigma_W^2 = \frac{\langle W^2 \rangle}{\langle W \rangle^2} - 1. \tag{4.66}$$

4.5.2 Quantum Nature of Radiation

As mentioned, there is another source of fluctuation, i.e., the quantum discrete nature of radiation. As a result of the Campbell's theorem [10], we know that for a Poisson photon statistics, the variance of photon number arising from this equals its expectation value. With both contribution from pointlike nature of electrons and quantum nature of radiation taken into account, the relative fluctuation of the radiation power or energy at a given frequency and a specific observation angle is

$$\frac{\text{Var}\,[P(\omega)]}{\langle P(\omega)\rangle^2} = \frac{1}{\langle N_{\text{ph}}(\omega)\rangle|_{\text{beam}}} + \frac{2}{N_e}\left(\frac{|\bar{b}(\omega)|^2 + \text{Re}\left[\bar{b}(2\omega)\bar{b}^2(-\omega)\right]}{|\bar{b}(\omega)|^4} - 2\right) + O\left(\frac{1}{N_e^2}\right),$$

(4.67)

in which

$$\langle N_{\text{ph}}(\omega)\rangle|_{\text{beam}} = \left[N_e + N_e(N_e - 1)|\bar{b}(\omega)|^2\right]\langle N_{\text{ph}}(\omega)\rangle|_{\text{point}}$$
$$\approx N_e^2\,|\bar{b}(\omega)|^2\,\langle N_{\text{ph}}(\omega)\rangle|_{\text{point}}$$

(4.68)

is the expected radiated photon number from the electron beam, and $\langle N_{\text{ph}}(\omega)\rangle|_{\text{point}}$ is the expected radiated photon number from a single electron. Note that to obtain a nonzero expected photon number $\langle N_{\text{ph}}(\omega)\rangle|_{\text{beam}}$, a finite frequency bandwidth is needed. Therefore, Eq. (4.67) actually applies to a finite frequency bandwidth close to ω where $\bar{b}(\omega)$ does not change much.

From Eq. (4.67), it is interesting to note that with the narrowing of the energy bandwidth acceptance, i.e., the decrease of $\langle N_{\text{ph}}(\omega)\rangle|_{\text{beam}}$, the contribution to the relative fluctuation from the quantum nature of radiation increases, while the contribution from the pointlike nature of electron does not change. This reflects the fact that one fluctuation is quantum, while the other is classical.

Note that in our interested case, $N_e\langle N_{\text{ph}}(\omega)\rangle|_{\text{point}}$ is usually much larger than 1, then the second term in Eq. (4.67) dominants. In other words, the fluctuation due to the pointlike nature of electrons dominants. Only when $N_e\langle N_{\text{ph}}(\omega)\rangle|_{\text{point}}$ is close to 1, will the first term become significant compared to the second term.

As the statistical property of the radiation embeds rich information about the electron beam, innovative beam diagnostics method can be envisioned by making use of this fact. One advantage of using radiation fluctuation in diagnostics is that it has a less stringent requirement on the calibration of detectors. Here we propose an experiment to measure the sub-ps bunch length accurately at a quasi-isochronous storage ring, for example the MLS, at a low beam current, by measuring and analyzing the fluctuation of the coherent THz radiation generated from the electron bunch. Equation (4.67) or some numerical code based on the analysis presented in this section will be the theoretical basis for the experimental proposal. In principle, we can also deduce the transverse distribution of electron beam by measuring the two-dimensional distribution of the radiation fluctuation. More novel beam diagnostics methods may be invented for SSMB and future light sources by making use of the statistical property of radiation.

4.6 Example Calculations for Envisioned EUV SSMB

To summarize our investigations on the average and statistical property of SSMB radiation, here we present an example calculation for the envisioned EUV SSMB. Figure 4.9 is an example plot of the beam current and longitudinal form factor spectra of the envisioned EUV SSMB. In the envisioned example, the microbunch length is $\sigma_z \approx 3$ nm at the radiator where 13.5 nm coherent EUV radiation is generated, and these 3 nm microbunches are separated from each other with a distance of

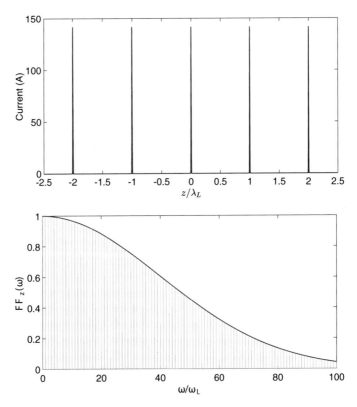

Fig. 4.9 An example plot of the beam current and longitudinal form factor spectrum of the microbunch train at the radiator in the envisioned EUV SSMB. Up: beam current of the 3 nm microbunch train separated by the modulation laser wavelength $\lambda_L = 1064$ nm. Bottom: longitudinal form factor $FF_z(\omega)$. The exponential decaying envelope corresponds to that of a single 3 nm Gaussian microbunch, and the green periodic delta function lines correspond to the periodic microbunch train in time domain. The desired radiation wavelength is $\lambda_0 = \frac{\lambda_L}{79} = 13.5$ nm

$\lambda_L = 1064$ nm $= 79 \times 13.5$ nm, which is the modulation laser wavelength. The radiator is assumed to be an undulator. The beam at the radiator can be round or flat depending on the lattice scheme, and its transverse size can range from a couple of μm to a couple of 10 μm. Note that a Gaussian-distributed current at the radiator is assumed in the plot. This is the case corresponding the usual longitudinal strong focusing SSMB. We remind the readers that the current at the radiator in the TLC-based bunch compression scheme, i.e., generalized longitudinal strong focusing, is actually non-Gaussian considering the nonlinear modulation waveform, as shown in Fig. 3.4. As our goal is to give the readers a picture of the radiation characteristics, here for simplicity we consider the case of a round beam. We remind the readers that the parameters used in this example EUV SSMB radiation calculation is for an illustration and is not optimized.

4.6.1 Average Property

First we present the result for the average property of the EUV radiation. The calculation is based on Eqs. (4.7), (4.11) and (4.22), and the result is shown in Fig. 4.10. The upper part of the figure shows the radiation energy spectrum. The lower part shows the spatial distributions of the radiation energy. The total radiation power is calculated according to

$$P = \frac{W}{\lambda_L/c}, \qquad (4.69)$$

where W is the total radiation energy loss of each microbunch. For the example radiator undulator parameters, corresponding to $\sigma_\perp = 5, 10, 20$ μm, the total radiation power are 92 kW, 14 kW, 3.5 kW, respectively. As a reference, the radiation power calculated based on Eq. (4.45) for these three transverse beam sizes are 4 kW, 3.6 kW and 2.6 kW, respectively. The reason Eq. (4.45) gives a smaller value than the overall power as explained is that it does not take into account the red-shifted part of the radiation. Therefore, Eq. (4.45) can be used to evaluate the lower bound of the radiation power from SSMB, once the parameters set of electron beam and radiator undulator is given. It can be seen that generally, kW-level EUV radiation power can be straightforwardly anticipated from a 3 nm microbunch train with an average beam current of 1 A. Note that for simplicity, in this example calculation, the filling factor of microbunches in the ring is assumed to be 100%, i.e., one microbunch per modulation laser wavelength. Then 1 A average current corresponds to the number of electrons per microbunch $N_e = \frac{I_A \lambda_L/c}{e} = 2.2 \times 10^4$, if the modulation laser wavelength is $\lambda_L = 1064$ nm.

Another important observation is that the spectral and spatial distribution of SSMB radiation depends strongly on the transverse size of the electron beam. A large transverse size results in a decrease of the overall radiation power, and also makes the radiation more narrowbanded and collimated in the forward direction. This is an important observation drawn from our investigation on the generalized transverse

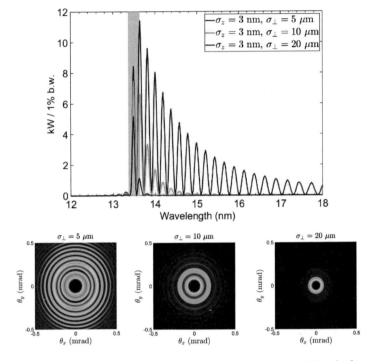

Fig. 4.10 An example EUV SSMB radiation calculation with a microbunch length of $\sigma_z = 3$ nm and different transverse sizes $\sigma_\perp$. The upper part shows the energy spectrum. Corresponding to $\sigma_\perp = 5, 10, 20$ μm, the total radiation power are 92 kW, 14 kW, 3.5 kW, respectively. The shaded area corresponds to wavelength of $13.5 \pm \frac{13.5}{100}$ nm. The bottom part shows spatial distribution of radiation energy. From left to right: $\sigma_\perp = 5, 10, 20$ μm. Parameters used for the calculation: $E_0 = 400$ MeV, $I_{avg} = 1$ A, $\lambda_L = 1064$ nm, $\lambda_r = \frac{\lambda_L}{79} = 13.5$ nm, $\lambda_u = 1$ cm, $K = 1.14$, $N_u = 2 \times 79$

form factor. Using the example parameters, i.e., $E_0 = 400$ MeV, $\lambda_L = 1064$ nm, $\lambda_0 = \frac{\lambda_L}{79} = 13.5$ nm, $\lambda_u = 1$ cm, $N_u = 2 \times 79$, $K = 1.14$, if $\sigma_\perp = 10$ μm, then according to Eqs. (4.39) and (4.40), the relative bandwidth and opening angle due to the transverse form factor can be calculated to be

$$\left.\frac{\Delta\omega_{e^{-1}}}{\omega_0}\right|_\perp \approx 1.7\%,$$

$$\left.\theta_{e^{-1}}\right|_\perp \approx 0.21 \text{ mrad.}$$

(4.70)

which is in agreement with the result presented in Fig. 4.10.

Since $N_u = 2 \times 79$ is used in the calculation, which means the radiation slippage length is twice the modulation laser wavelength, the energy spectrum and spatial distributions presented in Fig. 4.10 is for that of two neighboring microbunches. The peaks in the energy spectrum, and the interference rings in the radiation spatial distribution as explained is due to the macro longitudinal form factor $\left(\frac{\sin\left(N_b \frac{\omega}{c} \frac{\lambda_L}{2} \right)}{N_b \sin\left(\frac{\omega}{c} \frac{\lambda_L}{2} \right)} \right)^2$ of multiple microbunches, with N_b the number of microbunches. There is a one-to-one correspondence between the peaks in the energy spectrum and the interference rings in radiation energy spatial distribution. The reason for the appearance of such peaks and rings is that our radiation wavelength is a high harmonic of the distance between the neighboring microbunches.

As a result of the high-power and narrowband feature of the SSMB radiation, a high EUV photon flux of $10^{15} \sim 10^{16}$ phs/s within a 0.1 meV bandwidth can be obtained, if we can realize an EUV power of $\gtrsim 1$ kW per 1% b.w. as shown in Fig. 4.10. We remind the readers that the radiation waveform of SSMB is actually a CW or quasi-CW one, if induction linac is used as the energy compensation system and the microbunches occupy the ring with a large filling factor, as assumed in the example calculation. This kind of CW or quasi-CW narrowband photon source is favored in ARPES to minimize the space charge-induced energy shift, spectral broadening and distortion of photoelectrons in a pulsed photon source-based ARPES [11]. Therefore, the high photon flux within a narrow bandwith, together with its CW or quasi-CW waveform, makes SSMB a promising light source for ultrahigh-resolution ARPES. Such a powerful tool may have profound impact on fundamental physics research, for example to probe the energy gap distribution and electronic states of superconducting materials like the magic-angle graphene [12].

4.6.2 Statistical Property

Now we present the result for the statistical property of the radiation. For the case of a Gaussian bunch with $\sigma_z = 3$ nm and $N_e = 2.2 \times 10^4$, from Eq. (4.61) we know that the relative fluctuation of the turn-by-turn or microbunch-by-microbunch on-axis 13.5 nm coherent radiation power will be around 2%.

Figure 4.11 gives an example plot for the longitudinal form factor spectrum of three possible realizations of such a Gaussian microbunch. As can be seen, the spectrum is noisy mainly in the high-frequency or short-wavelength range. Since our EUV radiation is mainly at the wavelength close to 13.5 nm, and the longitudinal form factor close to this frequency fluctuates together from turn to turn, or bunch to bunch. As shown in Fig. 4.8 and discussed before, for the envisioned EUV SSMB, the transverse form factor fluctuation is much smaller than that of the longitudinal form factor. So the overall radiation power fluctuation is also about 2% as analyzed above. This fluctuation is also the fluctuation of microbunch center motion induced by the coherent radiation. A small fluctuation as it is, its beam dynamics effects need further study.

Note that this 2% fluctuation of radiation power should have negligible impact for the application in EUV lithography, since the revolution frequency of the microbunch in the ring is rather high (MHz), let alone if we consider that there is actually a

Fig. 4.11 The spectrum of
the longitudinal form factor
of three possible realizations
of a Gaussian microbunch
length of $\sigma_z = 3$ nm and
$N_e = 2.2 \times 10^4$. The shaded
area corresponds to
wavelength of $13.5 \pm \frac{13.5}{100}$ nm

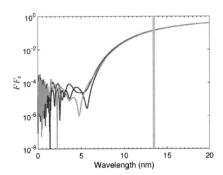

microbunch each modulation laser wavelength and the radiation waveform is CW or quasi-CW.

4.6.3   Discussions

To resolve possible concerns of readers on the validity of the short bunch length and high average current used in the example calculation, here we present a short discussion on the related single-particle and collective effects in SSMB. We recognize that realizing a steady-state bunch length as short as nanometer level in an electron storage ring is non-trivial. Both global and local momentum compaction should be minimized to confine the longitudinal beta function, therefore the longitudinal emittance, in an electron storage ring as analyzed in Sect. 2.1. By invoking this principle in the lattice design, a bunch length as short as tens of nanometer can be realized in a storage ring, with a momentum compaction factor of 1×10^{-6} and modulation laser power of 1 MW. 1 MW intra-cavity power is the state-of-art level of present optical enhancement cavity technology. Therefore, to realize nanometer bunch length at the radiator, we need to compress the bunch further. There are two scenarios being actively studied by us, namely the longitudinal strong focusing scheme and transverse-longitudinal coupling scheme. The longitudinal strong focusing scheme is similar to its transverse counterpart which is the basis of modern particle accelerators [13, 14]. In such a scheme, the longitudinal beta function and therefore the bunch length is strongly focused at the radiator, and the synchrotron tune of the beam in the ring can be at the level of 1, as analyzed in Sect. 2.1.6 and Ref. [15]. Although nanometer bunch length can be realized, this scheme requires a large modulation laser power (20 MW level if 270 nm laser is used), causing the optical cavity can work only in the pulsed laser mode and the average output radiation power is thus limited. To lower the modulation laser power, the transverse-longitudinal coupling scheme is thus applied in a clever way by taking advantage of the fact that the vertical emittance in a planar storage ring is rather small, as analyzed in Sect. 3.1. We refer to this turn-by-turn transverse-longitudinal coupling-based bunch compression scheme as the generalized longitudinal strong focusing [16], in which the phase space manipulation is 4D or 6D, in contrast to the conventional longitudinal strong focusing where the phase space manipulation is 2D. This generalized longitudinal strong focusing scheme can relax the modulation laser power, but its nonlinear dynamics optimization is a challenging task which we are trying to tackle.

Concerning the high average current, there are two collective effects of special importance, namely the intrabeam scattering (IBS) and coherent synchrotron radiation (CSR). IBS will affect the equilibrium emittance and thus can have an impact on the radiation power and also the modulation laser power in the generalized longitudinal strong focusing scheme. The IBS effect in an SSMB ring thus needs careful optimization and the operation beam energy is also mainly determined by IBS. CSR is the reason why SSMB can provide powerful radiation. On the other hand, CSR is also the effect which sets the upper limit of the stable beam current. In Ref. [17], there is some preliminary evaluation of the stable beam current for SSMB based on the 1D model of CSR-driven microwave instability. The investigation in this dissertation implies that the transverse dimension of the electron beam can have large impact on the coherent radiation in SSMB. In addition, the bunch lengthening from the transverse emittance in an SSMB storage ring can easily dominate the bunch length at many dispersive places of the ring, as the transverse size of microbunches is much larger than its longitudinal length. This bunch lengthening might be helpful in mitigating unwanted CSR. The 3D effect of the coherent radiation is expected to be also helpful in improving the stable beam current. With these beneficial arguments in mind, we realize that CSR in SSMB still deserves special attention. For example, the coherent radiation in the laser modulator could potentially drive single-pass and multi-pass collective instabilities in an SSMB storage ring [18, 19]. More in-depth study of collective effects in SSMB is ongoing.

References

1. Deng XJ, Zhang Y, Pan ZL, Li ZZ, Bian JH, Tsai C-Y, Li RK, Chao AW, Huang WH, Tang CX (2023) Average and statistical properties of coherent radiation from steady-state microbunching. J Synchrotron Radiat 30:35–50
2. Courant ED, Snyder HS (1958) Theory of the alternating-gradient synchrotron. Ann Phys (NY) 3(1):1–48
3. Jackson JD (1999) Classical electrodynamics
4. Chao AW (2020) Lectures on accelerator physics. World Scientific
5. Saldin EL, Schneidmiller EA, Yurkov MV (2005) A simple method for the determination of the structure of ultrashort relativistic electron bunches. Nucl Instrum Methods Phys Res A 539(3):499–526
6. Lobach I, Lebedev V, Nagaitsev S, Romanov A, Stancari G, Valishev A, Halavanau A, Huang Z, Kim K-J (2020) Statistical properties of spontaneous synchrotron radiation with arbitrary degree of coherence. Phys Rev Accel Beams 23:090703
7. Saldin E, Schneidmiller E, Yurkov M (1998) Statistical properties of radiation from vuv and x-ray free electron laser. Opt Commun 148(4):383–403

8. Lobach I, Nagaitsev S, Lebedev V, Romanov A, Stancari G, Valishev A, Halavanau A, Huang Z, Kim K-J (2021) Transverse beam emittance measurement by undulator radiation power noise. Phys Rev Lett 126:134802
9. Goodman JW (2015) Statistical optics. Wiley
10. Campbell N (1909) The study of discontinuous phenomena. Proc Cambridge Phil Soc 15:117–136
11. Zhou X, Wannberg B, Yang W, Brouet V, Sun Z, Douglas J, Dessau D, Hussain Z, Shen Z-X (2005) Space charge effect and mirror charge effect in photoemission spectroscopy. J Electron Spectrosc Relat Phenom 142(1):27–38
12. Cao Y, Fatemi V, Fang S, Watanabe K, Taniguchi T, Kaxiras E, Jarillo-Herrero P (2018) Unconventional superconductivity in magic-angle graphene superlattices. Nature 556(7699):43–50
13. Christofilos N (1950) Focusing system for ions and electrons. US Patent, vol 2, pp 736–799
14. Courant ED, Livingston MS, Snyder HS (1952) The strong-focusing synchroton-a new high energy accelerator. Phys Rev 88:1190–1196
15. Zhang Y (2022) Research on longitudinal strong focusing SSMB ring. PhD thesis, Tsinghua University, Beijing
16. Li Z, Deng X, Pan Z, Tang C, Chao A (2023) A generalized longitudinal strong focusing storage ring, Generalized longitudinal strong focusing in a steady-state microbunching storage ring, submitted
17. Chao A, Granados E, Huang X, Ratner D, Luo H-W (2016) High power radiation sources using the steady-state microbunching mechanism. In: Proceedings of the 7th international particle accelerator conference (IPAC'16), Busan, Korea, 2016. JACoW, Geneva, pp 1048–1053
18. Tsai C-Y, Chao AW, Jiao Y, Luo H-W, Ying M, Zhou Q (2021) Coherent-radiation-induced longitudinal single-pass beam breakup instability of a steady-state microbunch train in an undulator. Phys Rev Accel Beams 24:114401
19. Tsai C-Y (2022) Theoretical formulation of multiturn collective dynamics in a laser cavity modulator with comparison to robinson and high-gain free-electron laser instability. Phys Rev Accel Beams 25:064401

Chapter 5
SSMB Proof-of-Principle Experiments

To make SSMB a real option for future photon source facility, a crucial step is to experimentally demonstrate its working mechanism. In this chapter, we report the first successful experimental demonstration of the SSMB mechanism. Parts of the work presented have been published in Refs. [1, 2].

5.1 Strategy of the PoP Experiments

5.1.1 Three Stages of PoP Experiments

Considering the fact that it is a demanding task to realize SSMB directly in an existing machine, part of the reasons we have analyzed in previous chapters, among them the most fundamental one is the large quantum diffusion of bunch length in rings not optimized for SSMB, the SSMB PoP experiment has been divided into three stages as shown in Fig. 5.1. Some brief descriptions of the three stages are as follows.

- Phase I: a single-shot laser is fired to interact at the undulator with the electron beam stored in a quasi-isochronous ring. The modulated electron beam becomes microbunched at the same place of modulation after one complete revolution in the ring and this microbunching can preserve for several revolutions. By doing this experiment, we want to confirm that the optical phases, i.e., the longitudinal coordinates, of electrons can be correlated turn-by-turn in a sub-laser-wavelength precision. The realization of SSMB relies on this precise turn-by-turn phase correlation.
- Phase II: on the basis of Phase I, we replace the single-shot laser with a high-repetition phase-locked one to interact with the electrons at the undulator turn after turn. In this stage, we want to establish stable microbuckets and sustain the microbunching in the microbuckets to reach a quasi steady state.

© The Author(s) 2024
X. Deng, *Theoretical and Experimental Studies on Steady-State Microbunching*,
Springer Theses, https://doi.org/10.1007/978-981-99-5800-9_5

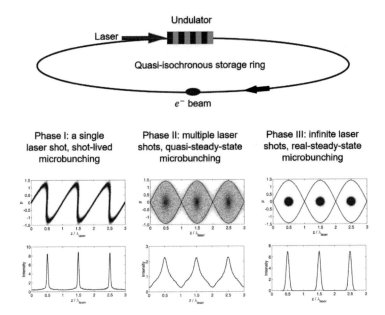

Fig. 5.1 Three stages of the SSMB PoP experiments: from single-shot to multiple shots to infinite shots laser pulse; from short-lived to quasi-steady-state to real steady-state microbunching

- Phase III: Phase II is very close to, but is still not, the final SSMB as a true SSMB means the balance of excitation and damping. However, the requirement of a true SSMB on the magnet lattice is demanding, especially the quantum diffusion of longitudinal coordinate z as analyzed in Sect. 2.1. Therefore, this final stage is more likely to be realized in a dedicated ring designed for SSMB, which is also one of the key ongoing tasks of the SSMB task force [3].

Below, we use PoP I, II, III to represent the three stages of the experiment. The key words of the three stage experiments are summarized as follows.

- PoP I: microbunching based on stored electron bunch, turn-by-turn phase correlation;
- PoP II: bounded motion in microbuckets, quasi steady state;
- PoP III: balance of diffusion and damping, real steady state.

These three stages each have their own significance and are all important for the SSMB development. Among them, Phase I is from 0 to 1, and is the most important one from conceptual viewpoint. Recently, we have successfully performed the PoP I and demonstrated the mechanism of SSMB at the Metrology Light Source (MLS) of Physikalisch-Technische Bundesanstalt (PTB) in Berlin [1, 2]. The experiment is a collaboration work of Tsinghua, Helmholtz-Zentrum Berlin (HZB) and PTB.

Table 5.1 Basic parameters of the MLS lattice

Parameter	Value	Description
E_0	50–630 MeV	Beam energy
C_0	48 m	Ring circumference
f_{RF}	500 MHz	RF frequency
V_{RF}	≤600 kV	RF voltage
η	3×10^{-2}	Phase slippage factor (standard user)
η	-2×10^{-5}	Phase slippage factor (SSMB experiment)
U_0	226 eV@250 MeV	Radiation loss per turn
J_s	1.95	Longitudinal damping partition
τ_δ	180 ms@250 MeV	Longitudinal radiation damping time
σ_δ	1.8×10^{-4}@250 MeV	Natural energy spread
σ_z	36 μm (120 fs)@250 MeV	Zero-current bunch length (SSMB experiment)
ν_x	3.18	Horizontal betatron tune
ν_y	2.23	Vertical betatron tune
ϵ_x	31 nm@250 MeV	Horizontal emittance
λ_L	1064 nm	Modulation laser wavelength
λ_u	125 mm	Undulator period length
N_u	32	Number of undulator periods
L_u	4 m	Undulator length
K	2.5	Undulator parameter

5.1.2 Metrology Light Source Storage Ring

The MLS is a storage ring optimized for quasi-isochronous operation [4–6], thus an appropriate testbed for SSMB physics investigation and PoP experiments. However, the partial phase slippage of the MLS is large as the bending angle of each dipole is large ($\frac{\pi}{4}$) and the dispersion magnitude inside the dipoles is also large, so it is not feasible to realize true SSMB, i.e., PoP III, directly at the MLS. Therefore, the SSMB PoP experiment has been divided into three stages as introduced just now, and PoP I and II are what we have performed and plan to conduct at the MLS. Some basic parameters of the MLS are shown in Table 5.1. The lattice optics of the MLS used in the SSMB PoP experiments are shown in Fig. 5.2.

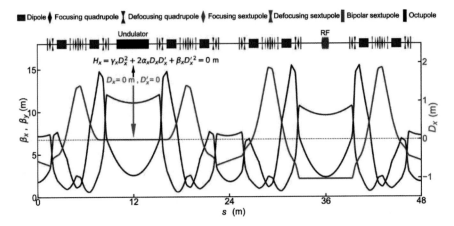

Fig. 5.2 The MLS quasi-isochronous magnet lattice used to generate microbunching. The magnet lattice and the key are shown at the top. The curves are the model horizontal (red) and vertical (blue) β-functions and the horizontal dispersion D_x (green). Operating parameters of the ring: beam energy, $E_0 = 250$ MeV; relative energy spread, $\sigma_\delta = 1.8 \times 10^{-4}$ (model); horizontal emittance, ϵ_x = 31 nm (model); horizontal betatron tune, $\nu_x = 3.18$ (model and measured); vertical betatron tune, $\nu_y = 2.23$ (model and measured); horizontal chromaticity, $\xi_x = -0.5$ (measured). Note that this optics is different from that used in Sect. 2.1.3 for the simulation of partial phase slippage effect. (Figure from Ref. [1])

5.2   PoP I: Turn-by-Turn Laser-Electron Phase Correlation

5.2.1   Experimental Setup

Figure 5.3 shows the schematic setup of the SSMB PoP I experiment. A horizontally polarized laser pulse (wavelength, $\lambda_L = 1064$ nm; pulse length, full-width at half-maximum, FWHM ≈ 10 ns; pulse energy, ≈ 50 mJ) is sent into a planar undulator (period, $\lambda_u = 125$ mm; total length, $L_u = 4$ m) to co-propagate with the electron bunches (energy, $E_0 = 250$ MeV) stored in the MLS storage ring (circumference, C_0 = 48 m). To maximize the laser-electron energy exchange, the undulator gap is chosen to satisfy the resonance condition $\lambda_s = \lambda_L$, where $\lambda_s = \frac{1+K^2/2}{2\gamma^2}\lambda_u$ is the central wavelength of the spontaneous undulator radiation, with $\gamma \propto E_0$ being the Lorentz factor and $K = \frac{eB_0}{m_e c k_u} = 0.934 \cdot B_0[\text{T}] \cdot \lambda_u[\text{cm}]$ being the dimensionless undulator parameter, determined by the undulator period and magnetic flux density. This laser-electron interaction induces a sinusoidal energy modulation pattern on the electron beam with a period of the laser wavelength. Because particles with different energies have slightly different revolution periods, after one revolution in the ring, the energy-modulated electrons shift longitudinally with respect to each other, clumping towards synchronous phases and forming microbunches. The formed microbunches can last several revolutions in the ring. The coherent undulator radiation generated

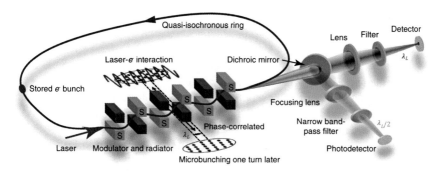

Fig. 5.3 Schematic of the experimental setup. The stored 250 MeV electron bunches are energy-modulated by a 1064 nm wavelength laser in an undulator, and become microbunched after one complete revolution in the 48 m circumference quasi-isochronous storage ring. This formed microbunching can then preserve for multiple turns in the ring. Each time the microbunching going through the undulator, narrowband coherent radiation will be generated. The undulator radiation is separated into the fundamental and second harmonics by dichroic mirrors, and sensitive photodiodes are used as the detectors. Narrow band-pass filters can be inserted in front of the photodetectors to pick out the narrowband coherent radiation generated from the microbunching. (Figure from Ref. [1])

from the microbunches, detected by a high-speed photodetector with a photodiode, confirms microbunching.

The symplectic longitudinal dynamics of the the above experiment processes can be modeled by

$$\begin{cases} \delta_1 = \delta_0 + A\sin(k_L z_0), \\ z_1 = z_0 - \eta C_0 \delta_1, \end{cases} \tag{5.1}$$

for the first revolution with laser modulation, and

$$\begin{cases} \delta_{m+1} = \delta_m, \\ z_{m+1} = z_m - \eta C_0 \delta_{m+1}, \end{cases} \tag{5.2}$$

for the later revolutions. This demonstration proves that the longitudinal dynamics described by the *one-turn map* Eq. (5.1) can be extrapolated from the RF wavelength (metre scale) to laser wavelength (micrometre scale) for a *stored* electron beam, thus validating the SSMB microbunching mechanism.

5.2.2 Physical Analysis of Microbunching Formation

5.2.2.1 Storage Ring

Operation energy The above models Eqs. (5.1) and (5.2), however, do not consider the non-symplectic, transverse-longitudinal coupling and nonlinear lattice dynamics, which all could lead to degradation of the microbunching. It turns out that the first non-symplectic dynamics we need to account for is the synchrotron radiation. As we know, when a relativistic electron is subjected to an acceleration normal to its velocity exerted by a bending magnet, it radiates electromagnetic energy. This radiation is characterized by the quantum nature of the photon emission process. The photon energy and emission place or time are both stochastic, giving rise to changes on particle energy (instantly) and the longitudinal coordinate z (non-instantly), as analyzed before in Sect. 2.1. Of special interest in the experiment is the root-mean-square (RMS) quantum diffusion of z in one turn d_z. According to Eq. (2.41), we have

$$d_z = \sqrt{\langle z^2 \rangle - \langle z \rangle^2} = \sqrt{\langle F^2 \rangle \langle \mathcal{N} \rangle \left\langle \frac{u^2}{E_0^2} \right\rangle}, \tag{5.3}$$

with $F(s_2, s_1) \equiv -\tilde{\eta}(s_2, s_1) C_0$. For the MLS quasi-isochronous magnet lattice used in the PoP experiment as shown in Fig. 5.2, d_z is as large as 260 nm at its standard operation energy of 630 MeV, deteriorating the sub-micrometre microbunching considerably. Therefore, the beam energy needs to be lowered to mitigate this diffusion, as $\sqrt{\langle \mathcal{N} \rangle \left\langle \frac{u^2}{E_0^2} \right\rangle} \propto \gamma^{2.5}$. At the same time, a lower beam energy gives a smaller energy spread and is also beneficial for microbunching, as the smearing from the natural uncorrelated energy spread becomes smaller. Nevertheless, the beam energy cannot be too low, otherwise the beam parameters and lifetime could be profoundly affected by scattering among particles [7–9]. An electron beam energy of 250 MeV is adopted in the experiment to balance these issues. At $E_0 = 250$ MeV, we have $d_z = 26$ nm and $\sigma_\delta = 1.8 \times 10^{-4}$.

Phase slippage factor Because the laser wavelength is much smaller than that of an RF wave, the phase slippage factor η needs to be ultrasmall. That is, the ring should be quasi-isochronous to allow turn-by-turn stabilization of the electron optical phases, i.e., the longitudinal coordinates, for particles with different energies. More quantitatively, the RMS spread of z in one turn that arises from the uncorrelated electron energy spread should be adequately smaller than the laser wavelength,

$$\Delta z_{ES} = |\eta C_0 \sigma_\delta| \leq \lambda_L / 2\pi. \tag{5.4}$$

To fulfill this requirement, the phase slippage factor of the MLS was lowered to $\eta \approx -2 \times 10^{-5}$, which is three orders of magnitude smaller than its standard value of 3×10^{-2}. By implementing these parameters, $\Delta z_{ES} = |\eta C_0 \sigma_\delta| \approx 0.17 \, \mu$m (about 0.6 fs), enabling the formation and preservation of sub-micrometre microbunching.

Such a quasi-isochronous magnet lattice is achieved by tailoring the horizontal dispersion functions D_x around the ring so that a particle with non-ideal energy travels part of the ring inwards and part of the ring outwards compared to the reference orbit, thus having a revolution period nearly the same as that of the ideal particle. The tailoring of D_x is accomplished by adjusting the (de)focusing strengths of the quadrupole magnet. The operation of the MLS as a quasi-isochronous ring also benefits from the optimization of the sextupole and octupole nonlinear magnet schemes to control the higher-order terms of the phase slippage [5, 6], which affect both the equilibrium beam distribution in the longitudinal phase space before the laser modulation and the succeeding microbunching evolution as analyzed in Sect. 2.2.1.

In the experiment, the value of the small phase slippage factor is quantified by measuring the electron orbit offsets while slightly adjusting the RF frequency up and down at a beam position monitor (BPM), where D_x is large. From the offsets and D_x, the phase slippage-dependent electron energy shifts caused by the RF frequency adjustment can be derived, as well as the phase slippage factor. The D_x value at the BPM is acquired using the same method in a reversible way, that is, based on a known phase slippage factor. This is done at a larger phase slippage factor, at which its value can be determined from its relation to the synchrotron oscillation frequency of the electron beam as given in Eq. (2.28). The synchrotron oscillation frequency can be measured accurately at a phase slippage factor such as -5×10^{-4}, and the model confirmed that the relative change of D_x at the highly dispersive BPM is small ($< 4\%$) when the phase slippage factor is reduced from -5×10^{-4} to the desired -2×10^{-5} by marginally changing the quadrupole magnet strengths.

Bunching factor With the electron beam evolved according to Eq. (5.1) for one revolution in the ring, the bunching factor at the n-th laser harmonic as analyzed in Sect. 2.2.1 is

$$b_n = J_n(nk_L\eta C_0 A)\exp\left[-\frac{(nk_L\eta C_0\sigma_\delta)^2}{2}\right], \tag{5.5}$$

where J_n is the n-th order Bessel function of the first kind. The coherent radiation power at the n-th harmonic is proportional to the bunching factor squared, $P_{n,\text{coh}} \propto |b_n|^2$. With the dynamics in the following turns modeled by Eq. (5.2), for the m-th revolution, we just need to replace the ηC_0 in Eq. (5.5) with $m\eta C_0$.

The maximum reachable bunching factor becomes larger with the decrease of η, given that the optimal A can always be realized. $\eta = -2 \times 10^{-5}$ is approximately the present lowest reachable value at the MLS, so here below we use this η for the analysis. As we will explain soon, our signal detection at first focuses on the second harmonic of the undulator radiation, since it is easier to mitigate the impact of the modulation laser on signal detection compared to the fundamental frequency. As shown in Fig. 5.4, given $\eta = -2 \times 10^{-5}$, a modulation strength of $A \approx 1.5\sigma_\delta$ results in the maximum bunching at the second harmonic. Correspondingly, given $A = 1.5\sigma_\delta$, the optimal η for the fundamental frequency and second harmonic bunching is a bit smaller than 2×10^{-5}. Note that the optimized conditions for the fundamental frequency and second harmonic bunching are different. Figure 5.5 shows the bunching factor evolution with respect to the revolution number, with $A = 1.5\sigma_\delta$ and $|\eta| = 2 \times 10^{-5}$. As can be seen, the second harmonic bunching can last only one turn, while the fundamental frequency bunching can last about three turns or even more if η becomes smaller, as also can be seen in the right part of Fig. 5.4. These expectations have also been confirmed in the experiment as will be presented soon.

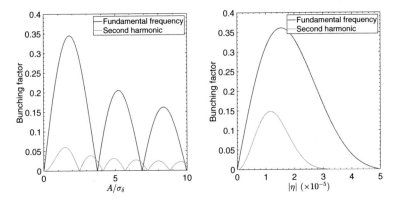

Fig. 5.4 Left: impact of the energy modulation strength ($A \propto \sqrt{P_L}$) on the bunching factor $|b_n|$ at the fundamental and second harmonic, with $\eta = -2 \times 10^{-5}$. Right: impact of the phase slippage factor η on the bunching factor $|b_n|$ at the fundamental and second harmonic, with $A = 1.5\sigma_\delta$

Fig. 5.5 The evolution of the bunching factor $|b_n|$ at the fundamental and second harmonic with respect to the revolution number, with $A = 1.5\sigma_\delta$ and $\eta = -2 \times 10^{-5}$

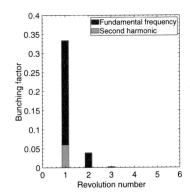

Chromatic $\mathcal{H}_x$ function and chromaticity ξ_x Apart from the longitudinal beam dynamics, the coupling of the particle betatron oscillation to the longitudinal dimension is also critical. The reason is based on the fact that the horizontal beam width at the undulator is about 600 μm (model value), three orders of magnitude larger than the sub-micrometre longitudinal structures that we aim to produce. Because the vertical emittance is much smaller than the horizontal one in a planar x-y uncoupled storage ring, in the following we consider only the impact of the horizontal betatron oscillation.

According to Eq. (3.13), for a periodic system, the RMS bunch lengthening of an electron beam longitudinal slice after m complete revolutions in the ring, due to betatron oscillation, is

$$\Delta z_{B,m} = 2\sqrt{\epsilon_x \mathcal{H}_x} \, |\sin(m\pi \nu_x)| . \tag{5.6}$$

Fig. 5.6 Influence of $\mathcal{H}_x$ at the undulator on the bunching factor one turn after laser modulation, with $\epsilon_x = 31$ nm (model) and $\nu_x = 3.18$

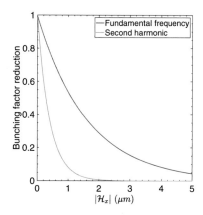

With this bunch lengthening taken into account, the bunching factor at the n-th laser harmonic after m revolutions will be

$$b_{n,m} = J_n(nk_L m\eta C_0 A)\exp\left[-\frac{(nk_L m\eta C_0 \sigma_\delta)^2 + \left(nk_L 2\sqrt{\epsilon_x \mathcal{H}_x}\,|\sin(m\pi \nu_x)|\right)^2}{2}\right].$$
(5.7)

The relative bunching factor reduction due to the non-zero $\mathcal{H}_x$ at the undulator can thus be defined as

$$R_{n,m}(\mathcal{H}_x) = \exp\left[-\frac{\left(nk_L 2\sqrt{\epsilon_x \mathcal{H}_x}\,|\sin(m\pi \nu_x)|\right)^2}{2}\right].$$
(5.8)

Putting in $\epsilon_x = 31$ nm (model) and $\nu_x = 3.18$ (model and measured), we need $\mathcal{H}_x \leq 0.8$ μm at the undulator to have $\Delta z_{B,1} \leq \lambda_L/2\pi$. Figure 5.6 shows the bunching factor reduction at the fundamental frequency and the second harmonic one turn after the laser modulation as a function of the $\mathcal{H}_x$ at the undulator. As can be seen, the second-harmonic bunching is even more sensitive to the $\mathcal{H}_x$ at the undulator. This stringent condition on $\mathcal{H}_x$ (note that $\mathcal{H}_x$ at other places of the ring is typically ≥ 0.1 m) is satisfied by fine-tuning the quadrupole magnet (de)focusing strengths to correct the dispersion D_x and dispersion angle D_x' at the undulator to the level of millimetre and 0.1 mrad, respectively (see Fig. 5.2).

In addition to the linear-order oscillating bunch lengthening, as discussed in Sect. 3.2, the betatron oscillation also produces an average path lengthening or shortening (second-order effect) described by the formula

$$\Delta C_B = -2\pi J_x \xi_x,$$
(5.9)

with ΔC_B being the average change of the particle recirculation path length, and $\xi_x = d\nu_x/d\delta$ being the horizontal chromaticity of the ring. Because different particles have different betatron oscillation amplitudes (actions), this effect results in a loss of synchronization between particles and degrades microbunching. Moreover, it broadens the equilibrium energy spread and distorts the beam from the Gaussian form before the laser modulation as investigated in Sect. 3.2, which also affects the microbunching. Therefore, the horizontal chromaticity should be small, to moderate its detrimental outcome, and simultaneously sufficient to suppress collective effects such as the head-tail instability [10]. As a consequence, a small negative chromaticity is used in the experiment.

5.2.2.2 Modulation Laser

Long-pulse laser A long-pulse laser (FWHM $\approx$ 10 ns) has been used to simplify the experiment by avoiding a dedicated laser-electron synchronization system, as the shot-to-shot laser timing jitter is $t_{jitter} \leq 1$ ns (RMS). According to Eq. (5.5) and Fig. 5.4, for a given phase slippage factor and harmonic number, there is an optimal laser-induced energy modulation amplitude A ($A \propto \sqrt{P_L}$ with P_L the laser power) that gives the maximum bunching factor. The laser used in the experiment (Beamtech Optronics Dawa-200) has multiple longitudinal modes, and its temporal profile has several peaks and fluctuates considerably from shot to shot (see Fig. 5.7). Therefore, the laser-induced electron energy modulation amplitudes are different from shot to shot and from bunch to bunch. When the modulation amplitude matches the phase slippage factor, the energy-modulated electrons are properly focused at synchronous phases, which gives optimal microbunching. For some of the shots, the laser intensity is higher or lower than the optimal value, and the electrons are then over-focused or under-focused, giving weaker microbunching and less coherent radiation. As we will see soon, this explains the large shot-to-shot fluctuation of the coherent amplified signals shown in Fig. 5.16c, e.

Power and Rayleigh length The electric field of a TEM00 mode Gaussian laser beam is [11]

$$E_x = E_{x0} e^{ik_L z - i\omega t + i\phi_0} \frac{1}{1 + i\frac{z}{Z_R}} \exp\left[i\frac{k_L Q}{2}(x^2 + y^2)\right],$$

$$E_z \approx -E_x x, \tag{5.10}$$

with $Z_R = \frac{\pi w_0^2}{\lambda_L}$ the Rayleigh length, w_0 the beam waist radius, and $Q = \frac{i}{Z_R(1+\frac{z}{Z_R})}$. The relation between E_{x0} and the laser peak power is given by

$$P_L = \frac{E_{x0}^2 Z_R \lambda_L}{4 Z_0}, \tag{5.11}$$

in which $Z_0 = 376.73$ Ω is the impedance of free space. The electron wiggles in an undulator according to

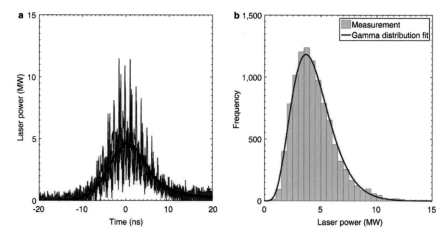

Fig. 5.7 Fluctuating temporal profiles of the multilongitudinal-mode laser. **a**, Temporal profiles of two example consecutive laser shots (red and blue) and the averaged waveform of 200 consecutive laser shots (black). **b**, Statistical distribution of the laser power at $t = 0$ ns in a for 10^4 consecutive laser shots, where the red curve is a gamma distribution fit. Laser: compact Nd:YAG Q-switched laser (Beamtech Optronics Dawa-200). Detector: ultrafast photodetectors (Alphas UPS-40-UVIR-D; rise time < 40 ps). Measurement system: digital oscilloscope (Teledyne LeCroy WM825Zi-B; bandwidth 25 GHz; sample rate 80 billion samples per second). (Figure from Ref. [1])

$$x(z) = \frac{K}{\gamma k_u} \sin(k_u z), \tag{5.12}$$

and the laser-electron exchange energy according to

$$\frac{dW}{dt} = ev_x E_x + ev_z E_z. \tag{5.13}$$

Assuming that the laser beam waist is in the middle of the undulator, and when $x, y \ll w(z)$, which is the case for SSMB PoP I, we drop the $\exp\left[i\frac{k_L Q}{2}(x^2 + y^2)\right]$ in the laser electric field. Further, when $Z_R \gg \lambda_u$, we can also drop the contribution from E_z on the energy modulation. The integrated modulation voltage induced by the laser in the planar undulator, normalized by the electron beam energy, is then [11]

$$\frac{eV_{\text{mod}}}{E_0} = \frac{e[JJ]K}{\gamma^2 mc^2} \sqrt{\frac{4P_L Z_0 Z_R}{\lambda_L}} \tan^{-1}\left(\frac{L_u}{2Z_R}\right), \tag{5.14}$$

in which $[JJ] = J_0(\chi) - J_1(\chi)$ and $\chi = \frac{K^2}{4+2K^2}$.

As can be seen from the above formula, when $\frac{L_u}{Z_R}$ is kept constant, then $V_{\text{mod}} \propto \sqrt{Z_R} \propto \sqrt{L_u}$. In our case, L_u is fixed, then as shown in Fig. 5.8, to maximize V_{mod} we need $\frac{Z_R}{L_u} = 0.359 \approx \frac{1}{3}$. On the other hand, the modulation strength does not depend on Z_R sensitively when Z_R is larger than this optimal value. To make the laser beam

Fig. 5.8 Integrated
modulation voltage of a laser
modulator as a function of
the ratio between laser
Rayleigh length and the
modulator undulator length

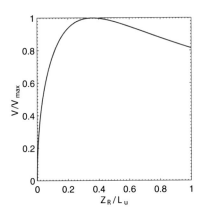

waist larger than the electron beam and thus induce the same energy modulation
on different electrons, a larger Rayleigh length might be in favored in our case.
For example if $Z_R = 2L_u$, then in order to induce an energy modulation depth of
$A = 1.5\sigma_\delta$, we can calculate that the laser power required is 430 kW. Considering the
non-ideal conditions and the fact that the laser used contains higher-order Gaussian
modes, one order of magnitude higher laser power might be required in the actual
case.

5.2.2.3   Microbunching Simulation

Based on the above parameters, we have conducted the simulation of microbunching
formation in the storage ring. The beam current and bunching factor one turn after
the laser modulation is shown in Fig. 5.9. As can be seen from the comparison with
Fig. 5.5, the simulation agrees with theory well.

5.2.3   *Microbunching Radiation Calculation*

Now we evaluate what radiation we can obtain from the formed microbunching.
The numerical calculation of incoherent and coherent undulation radiation shown
in this section are obtained using SPECTRA [13]. The beam energy and undulator
parameters used are those in our SSMB proof-of-principle experiment, i.e., $E_0 =$
250 MeV, $\lambda_L = \lambda_0 = 1064$ nm, $\lambda_u = 125$ mm, $K = 2.5$, $N_u = 32$. The results are
also used to compare with the theoretical formulas presented in Chap. 4. Note that the
numerical calculation of coherent radiation with a 3D charge distribution is usually
time-consuming. This is also one of the motivations for us in Chap. 4 to develop the
simplified analytical formulas with the main physics accounted for.

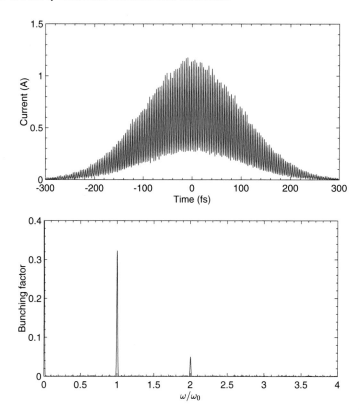

Fig. 5.9 Example current profile and bunching factor spectrum one turn after the laser modulation in SSMB PoP I, obtained from ELEGANT [12] tracking. Parameters used: $\sigma_t = 100$ fs, $\epsilon_x = 31$ nm, $\epsilon_y = \frac{1}{10}\epsilon_x$. 1×10^6 particles are simulated, meaning 160 fC for a one-to-one correspondence

The left part of Fig. 5.10 shows the incoherent undulator radiation flux of 10^6 electrons (0.16 pC) versus the opening angle of a circular aperture placed in the forward direction of electron traveling. As can be seen, with the increase of the aperture opening angle, the red-shifted part of the radiation grows. For the total flux, there are sharp spikes near the odd harmonics and no clear spikes near the even harmonics. This is due to the fact that there is no on-axis radiation at the even harmonics. Also note that with the change of the aperture opening angle, there are jumps in the flux at a specific frequency $n\omega_0$. This is due to the fact that the red-shifted radiation of higher harmonics $m > n$ can contribute to the flux at $\omega = n\omega_0$ when the aperture is large enough.

Now we calculate the coherent radiation of the laser modulation-induced microbunched beam, using an RMS bunch length of 100 fs ($\sigma_z = 30\,\mu$m). An example beam current and bunching factor spectrum of the laser modulation-induced microbunched beam are shown in Fig. 5.9. We remind the readers that the bunch length in the real machine is typically longer than that used in the calculation here.

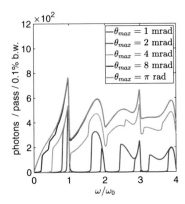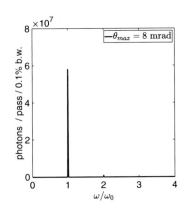

Fig. 5.10 Left: incoherent radiation photon flux of 10^6 electrons within a circular aperture placed in the forward direction, whose largest opening angles are θ_{max}. Right: coherent radiation photon flux of 10^6 electrons within the circular aperture whose largest opening angle $\theta_{max} = 8$ mrad. The beam current and bunching factor spectrum used in the calculation are shown in Fig. 5.9. Other related parameters: $E_0 = 250$ MeV, $\lambda_0 = 1064$ nm, $\lambda_u = 125$ mm, $K = 2.5$, $N_u = 32$

Therefore, the coherent radiation is even more narrowbanded than presented in the example calculation in this section.

At first, we ignore the influence of electron beam's transverse dimension, i.e., a thread beam is assumed. The coherent radiation spectrum of 10^6 electrons is shown in the right part of Fig. 5.10. As can be seen, there is narrowband coherent radiation only at the modulation laser harmonics, which fits with expectation due to the fact that there is only notable bunching factor at the laser harmonics. More closer look of the first two harmonics versus the aperture opening angle are shown in the upper part of Fig. 5.11. As can be seen, indeed the relative bandwidth of the coherent radiation at the fundamental frequency and second harmonic are 1% and 0.5%, respectively, agreeing with the values calculated from Eq. (4.21). In addition, the flux of the fundamental mode $H = 1$ at the fundamental frequency agrees reasonably well with that according to Eq. (4.48), i.e., $\mathcal{F}_1(\omega = \omega_0, \sigma_\perp = 0 \ \mu m) = 3.4 \times 10^7$ (photons/pass/0.1% b.w.). In other words, the amplification factor of the flux at $\omega = H\omega_0$ is indeed $N_e^2 |b_{z,H}|^2$ when $\sigma_\perp = 0 \ \mu m$. Also note that the jumps of the flux with the change of aperture opening angle as we commented just now.

Now we investigate the impact of transverse electron beam sizes on the coherent radiation. As can be seen from the middle and bottom parts of Fig. 5.11, which correspond to a transverse electron beam size of 100 μm and 400 μm, respectively, and the comparison with the upper part, the transverse sizes of the electron beam suppress the coherent radiation. And the calculated fluxes at the fundamental frequency agrees well with those predicted by Eq. (4.49), i.e., $\mathcal{F}_1(\omega = \omega_0, \sigma_\perp = 100 \ \mu m) = 3.1 \times 10^7$ (photons/pass/0.1% b.w.) and $\mathcal{F}_1(\omega = \omega_0, \sigma_\perp = 400 \ \mu m) = 1.6 \times 10^7$ (photons/pass/0.1% b.w.). The suppression from transverse beam size is even more significant at the higher harmonics and the suppression factors agree with those predicted according to the transverse form factor Eq. (4.32). Also note that

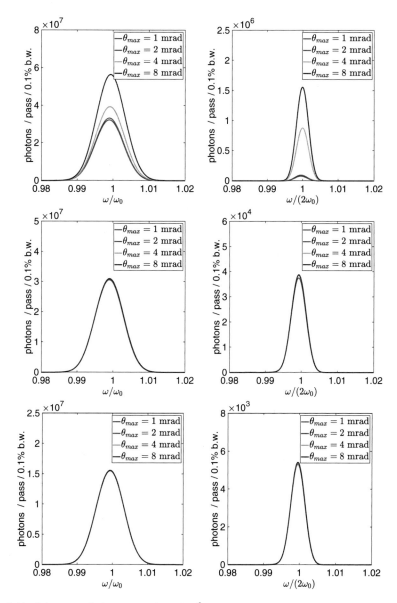

Fig. 5.11 Coherent radiation photon flux of 10^6 electrons versus opening angle θ_{max} of the circular aperture for the first two harmonics, with $\sigma_\perp = 0$ µm (up), $\sigma_\perp = 100$ µm (middle) and $\sigma_\perp = 400$ µm (bottom), respectively. The beam current and bunching factor spectrum used in the calculation are shown in Fig. 5.9. Other related parameters: $E_0 = 250$ MeV, $\lambda_0 = 1064$ nm, $\lambda_u = 125$ mm, $K = 2.5$, $N_u = 32$

Fig. 5.12 Total radiation power as a function of the observation time gathered within a circular aperture with $\theta_{max} = 1$ mrad, for $\sigma_\perp = 0, 100, 400$ μm, respectively. The beam current and bunching factor spectrum used in the calculation are shown in Fig. 5.9. Other related parameters: $E_0 = 250$ MeV, $\lambda_0 = 1064$ nm, $\lambda_u = 125$ mm, $K = 2.5$, $N_u = 32$

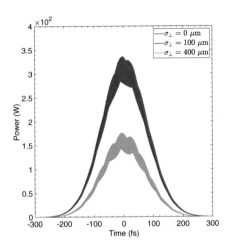

different from that of incoherent radiation, when $\sigma_\perp = 100$ μm or 400 μm, there is no visible jump of the flux with the aperture opening angle θ_{max} grown from 1 mrad to 8 mrad. This is because that the off-axis red-shifted coherent radiation of higher modes are suppressed now.

We have also confirmed our derivation of the coherent radiation power by comparing it with simulation. As shown in Fig. 5.12, the calculated peak powers of coherent radiation with different transverse electron beam sizes also agree well with the theoretical predictions from Eq. (4.46), i.e., $P_{1,peak}(\sigma_\perp = 0$ μm$) = 363$ W, $P_{1,peak}(\sigma_\perp = 100$ μm$) = 332$ W, $P_{1,peak}(\sigma_\perp = 400$ μm$) = 168$ W.

From the calculation and analysis, we know that the coherent radiation from the formed microbunching is mainly at the fundamental frequency and second harmonic of the modulation laser, and in the forward direction. The coherent radiation is narrowbanded, and stronger than the incoherent radiation. These calculations and observation are the basis for our signal detection scheme.

5.2.4 Signal Detection and Evaluation

After investigating the microbunching formation beam dynamics and radiation characteristics of the formed microbunching in the above sections, now we consider how we can measure and evaluate the signals.

Measurement and evaluation of bunch charge The bunch-by-bunch charge (current) in the experiment is measured by a single-bunch current monitor [14], which analyses the electron beam-induced RF signals from a set of four stripline electrodes (3 GHz bandwidth). To minimize the influence of neighbouring bunches on the signal, the pulse response of the electrodes is reshaped by a 500-MHz low-pass filter. The current calibration of the monitor is conducted using a parametric current

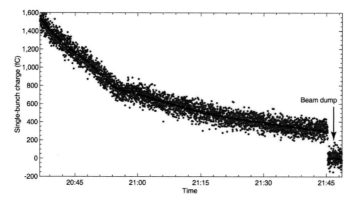

Fig. 5.13 Evaluation of bunch charge based on the stripe line measurement. Blue dots are the measurement results with the systematic offset subtracted and the red curve is a fit by the sum of two exponential functions, $Q(t) = Q_1 \exp(-t/\tau_1) + Q_2 \exp(-t/\tau_2)$, performed at different time intervals, with the fit results connected by a smoothed line. (Figure from Ref. [1])

transformer [4] at higher current, and the linearity of the system at lower current is checked with the signal of the photodiode illuminated by synchrotron radiation. During the current decay in the experiment, one data point of the result given by the monitor is saved every second for each individual bunch. The averaged measurement of ten unfilled bunches preceding the homogeneous filled bunches (10 ns time gap in between) is used as the systematic offset. To smooth the measurement noise and at the same time account for the change of the beam lifetime, the time evolution of the offset-removed data points is then fitted by the sum of two exponential functions, $Q(t) = Q_1 \exp(-t/\tau_1) + Q_2 \exp(-t/\tau_2)$, at different time intervals, with the fit results connected smoothly. One example evaluation of the bunch charge measurement result is presented in Fig. 5.13. Based on the evaluated data, we obtain a linear bunch-charge dependence of the broadband incoherent signal that is detected by the photodetector without the 3-nm-bandwidth band-pass filter, as shown in Fig. 5.14, confirming the reliability of the bunch-charge measurement and evaluation method.

If the ring works in single-bunch mode, there is a more direct and accurate method of bunch charge measurement based on the measurement of synchrotron radiation strength using a photodiode. Both methods have been used in our experiments.

Detection and evaluation of undulator radiation The long-pulse laser (FWHM ≈ 10 ns) is used to simplify the experiment by avoiding a dedicated laser-electron synchronization system, given that the shot-to-shot laser timing jitter is $t_{\text{jitter}} \leq 1$ ns (RMS). However, the photodetector (Femto HSPR-X-I-1G4-SI; rise/fall time, 250 ps) becomes saturated and even damaged by the powerful laser (Beamtech Optronics Dawa-200) if it is placed in the path of the laser. To address this issue, the undulator radiation is separated into the fundamental and second harmonics by appropriate dichroic mirrors (Thorlabs Harmonic Beamsplitters HBSY21/22), as shown in Fig. 5.3, and the signal detection at first focuses on the second harmonic with the

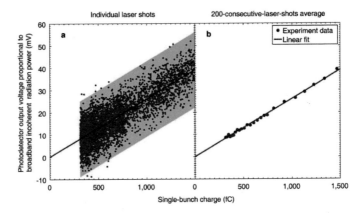

Fig. 5.14 Linear dependence of the broadband incoherent undulator radiation on the bunch charge. **a**, Results corresponding to individual laser shots; the shading (light red) represents 3σ of the detection noise. **b**, The result after 200-consecutive-laser-shot averaging. The blue dots are the experimental data of a bunch not modulated by the laser and the red curves are linear fits. (Figure from Ref. [1])

wavelength centred at 532 nm. The photodetector output voltage, which is proportional to the radiation power, is then measured by a digital oscilloscope (Tektronix MSO64:6-BW-4000; bandwidth, 4 GHz; sample rate, 25 billion samples per second). Later we will also present the result of the 1064 nm radiation by implementing Pockels cells to block the modulation laser and let pass the radiation in the following turns.

An example radiation waveform of the second harmonic is shown in Fig. 5.16. To avoid the impact of the signal waveform offset caused by stray laser light, the data analysis takes the peak-to-peak value of the photodetector output voltage as a measure of the radiation power. The coherent radiation power that corresponds to each individual laser shot, obtained during a time interval with a decaying beam current, is presented in Fig. 5.15a, where the modest contribution on the measured quantity from the small amount of incoherent radiation transmitted through the 3-nm-bandwidth band-pass filter has been eliminated. As can be seen, the coherent signal fluctuates considerably from shot to shot. This is attributable to the shot-to-shot fluctuation of the laser intensity profile (see Fig. 5.7) and the measurement noise. Despite the fluctuation, quadratic functions fit reasonably to the lower and upper bounds of the data points, which correspond to the cases of minimum and maximum bunching factors induced by the fluctuating laser, respectively. When performing the fits, we took into account that the measured quantity is the real radiation signal convoluted with the detection noise. The impact of this noise, obtained by analysing the measurement result of the unfilled bunches, on the bounds of the measured data points is visualized as shading in Fig. 5.15a for the coherent signal and in Fig. 5.14a for the incoherent signal. To smooth this shot-to-shot fluctuation, a 200-consecutive-laser-shot averaging is conducted and the results are presented in Figs. 5.14b and 5.15b

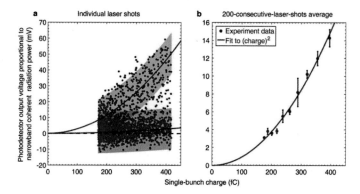

Fig. 5.15 Quadratic dependence of the narrowband coherent undulator radiation generated from microbunching on the bunch charge. **a**, Results corresponding to individual laser shots; the shading (light red and grey) represents 3σ of the detection noise. **b**, The result after 200-consecutive-laser-shot averaging; the plot is the same as Fig. 5.17 and is presented again here for comparison with **a** and with the incoherent signal in Fig. 5.14. The blue dots represent the experimental data and the red curves are quadratic fits. (Figure from Ref. [1])

for the narrowband coherent and broadband incoherent signals, where a quadratic and a linear fit have been performed, respectively.

5.2.5   Experimental Results

5.2.5.1   Second Harmonic Radiation

Figure 5.16 shows the typical measurement results of the second-harmonic undulator radiation emitted from a homogeneous stored bunch train, with a charge of about 1 pC per bunch and a time spacing of 2 ns, supplied by the 500 MHz RF cavity at the MLS. The spikes in the waveforms are the signals of different bunches. The left and right panels show the results corresponding to 2 and 40 consecutive laser shots, respectively. To smooth the measurement noise and signal fluctuation, the waveforms in the right panels have been averaged. Figure 5.16a, b shows the signals one turn before the laser shot, which correspond to the incoherent radiation and reflect the homogeneous bunch filling pattern. Figure 5.16c, d shows the radiation one turn after the laser shot, from the same bunches as those in Fig. 5.16a, b. The five larger spikes at the centre correspond to the bunches modulated by the laser. The enhanced signals of these five spikes indicate the formation of microbunching and the generation of coherent radiation from the laser-modulated bunches.

The laser used in the experiment has multiple longitudinal modes, and its temporal profile has several peaks and fluctuates considerably from shot to shot (see Fig. 5.7). Therefore, the laser-induced electron energy modulation amplitudes are different from shot to shot and from bunch to bunch. When the modulation amplitude matches

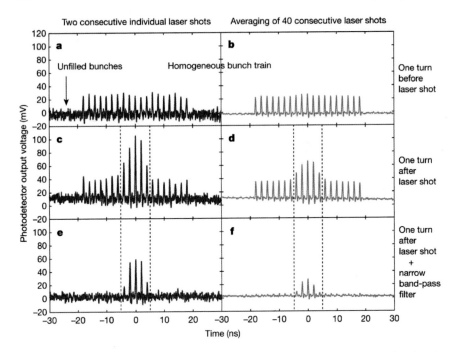

Fig. 5.16 Waveforms of the undulator radiation produced from a homogeneous stored bunch train.
a, b, Radiation one turn before the laser shot. The photodetector output voltage is proportional to
the radiation power. **c, d,** Radiation one turn after the laser shot, from the same bunches as those of **a**
and **b**, where the central five bunches are modulated by the laser pulse. The offset and general slight
decreasing trend of the waveforms are due to the photodetector being saturated by stray light from
the modulation laser one revolution before and not having completely recovered. **e, f,** Radiation
one turn after the laser shot, obtained with a narrow band-pass filter (centre wavelength, 532 nm;
bandwidth, 3 nm FWHM) placed in front of the photodetector, with bunch filling and charge similar
to those in **a** to **d**. (Figure from Ref. [1])

the phase slippage factor, the energy-modulated electrons are properly focused at
synchronous phases, which gives optimal microbunching. For some of the shots,
the laser intensity is higher or lower than the optimal value, and the electrons are
then over-focused or under-focused, giving weaker microbunching and less coherent
radiation. This explains the shot-to-shot fluctuation of the coherent amplified signals
shown in Fig. 5.16c, e.

As analyzed before, the microbunching coherent radiation is much narrowbanded
compared to the incoherent radiation. To confirm that the amplified radiation is due to
microbunching, we tested this narrowband feature of the coherent radiation. A band-
pass filter (Thorlabs FL532-3; centre wavelength, 532 nm; bandwidth, 3 nm FWHM)
was inserted in front of the detector. The radiation one turn after the laser shot is
shown in Fig. 5.16e, f, which was obtained with a bunch filling and charge similar to
that of Fig. 5.16c, d. From the comparison between Fig. 5.16e and c (Fig. 5.16f and
d), we can see that the broadband incoherent signals are nearly completely blocked

by the filter, whereas the amplified part is not affected much, confirming that the amplification is the narrowband coherent radiation generated by the microbunches.

Finally, we investigated the dependence of the coherent radiation on the bunch charge. To mitigate collective effects such as intrabeam scattering and head-tail instability, which could change the electron beam parameters, this investigation was conducted at low beam current, and the coherent signal was optimized by fine-tuning the machine to ensure a sufficient signal-to-noise ratio. Because the longitudinal radiation damping time in the experiment was 180 ms, we operated the laser at 1.25 Hz repetition rate to ensure that the electron bunches had time to recover their equilibrium parameters before each laser shot. The 3-nm-bandwidth band-pass filter was inserted to block the incoherent radiation, and the coherent signal corresponding to each individual laser shot was saved, with the beam current decaying naturally until the signal was at the detection noise level. The measurement results of the bunch closest to the laser temporal centre ($t = 0$ ns in Fig. 5.16) are used for quantitative analysis as introduced above. To lessen the impact of the laser temporal profile fluctuation and measurement noise, a 200-consecutive-laser-shot averaging is performed to obtain the data point for each bunch charge. The coherent undulator radiation power versus the single-bunch charge is shown in Fig. 5.17, where a quadratic function fits well to the experiment data. The quadratic bunch charge dependence, together with the narrowband feature of the coherent radiation, demonstrates unequivocally the formation of microbunching.

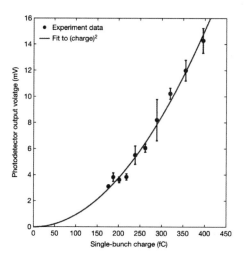

Fig. 5.17 Quadratic dependence of the coherent undulator radiation generated from microbunching on the bunch charge. The blue dots represent the experimental data and the red curve is a quadratic fit. Each data point represents the averaged result of 200 consecutive laser shots. The error bars denote the standard deviation of the averaged results when the averaging time window shifts for ± 100 consecutive laser shots from the corresponding data point. (Figure from Ref. [1])

5.2.5.2 Fundamental Frequency Radiation

The above deciding experimental results were obtained in the year of 2020, and have been published in Ref. [1]. After that, there are two main upgrades on the experimental setup. First, the multi-longitudinal-mode laser has been replaced by a single-longitudinal-mode one (Amplitude Surelite I-10). Second, Pockel Cells have been installed along the signal detection optical path to block the modulation laser and let pass the radiation in later revolutions, thus allowing the detection of the fundamental frequency radiation [15]. As shown in our analysis, we expect that the coherent radiation at the fundamental frequency is much stronger than that at the second harmonic and the microbunching can last multiple turns. These expectations have been confirmed in our following experimental investigations at the MLS [2].

Figure 5.18 shows the typical experimental results of the multi-turn coherent radiation at the fundamental frequency. A narrowband-pass filter has been inserted to select the narrowband coherent radiation. Figure 5.19 is the more quantitative data analysis of the signal of the first three turns after each laser shots. The bunch charge has now also been obtained in a more accurate way by using the incoherent synchrotron radiation signal measured by a photodiode. Several important observations are in order concerning the experiment results:

- First, signals of all three turns have shown nice quadratic bunch charge fits, confirming again the formation of microbunching and coherent radiation generation from it.
- Second, we mentioned in the above section and also in Ref. [1] that the huge shot-to-shot coherent radiation signal fluctuation obtained before the upgrades is mainly due to the laser profile fluctuation arising from its multi-longitudinal-mode nature. This argument has also been confirmed by the results shown in Fig. 5.19. From the comparison of Figs. 5.15a and 5.19, we can see that the shot-to-shot coherent signal with the present single-longitudinal-mode laser is much more stable.

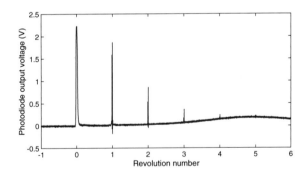

Fig. 5.18 Raw data of the multi-turn microbunching preservation experiment result. The signal is for the fundamental-mode undulator radiation, i.e., 1064 nm. (Refer to J. Feikes' talk in IPAC2021 and Ref. [2] for more details)

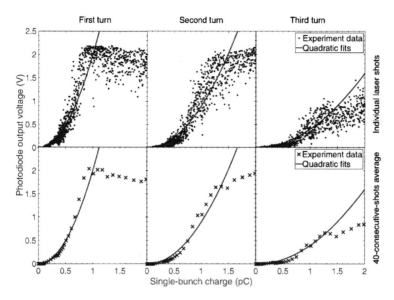

Fig. 5.19 The bunch-charge scaling of the coherent undulator radiation signals of the first three turns after the laser shots. Up: each data point corresponds to the radiation signal strength after each single laser shot. Bottom: average of the radiation signal from 40 consecutive laser shots. The saturation level of the detector is about 2 V

- Third, the signal deviates from the quadratic scaling at high current and starts to saturates about 1.3 pC, which we believe is due to the influence of collective effects.

More in-depth investigations on the multi-turn microbunching is still ongoing and will be reported in the future [2].

5.2.6 Summary

In conclusion, we have demonstrated the mechanism of SSMB for the first time in a real machine. This demonstration represents the first milestone towards the implementation of an SSMB-based high-repetition, high-power photon source. As great as the experimental results are, to avoid confusion, here we make clear that here we do not report an actual demonstration of SSMB, but rather a demonstration of the mechanism by which SSMB will eventually be attained. First, the formation of microbunching after one *complete* revolution of a laser-modulated bunch in a quasi-isochronous ring and the maintenance of microbunching for multiple turns demonstrate the viability of a turn-by-turn electron optical phase correlation with a precision of sub-laser wavelength. Second, this microbunching is produced on the *stored* electron bunch, the equilibrium parameters and distribution of which before the laser modulation are defined by the same storage ring as a whole. The combination of these two crucial factors establishes a closed loop to support the realization of SSMB, provided that a phase-locked laser interacts with the electrons turn by turn.

5.3 PoP II: Quasi-steady-State Microbunching

On the basis of the PoP I, the next step is to replace the single-shot laser by a high-repetition phase-locked one to interact with the electrons turn-by-turn. By doing so, we want to form stable microbuckets to constrain the microbunching in it to reach a quasi steady state. This is the SSMB PoP II as introduced in the beginning of this chapter.

5.3.1 Phase-Mixing in Buckets

To reach a quasi steady state, the particles need to do synchrotron oscillations to reach phase mixing in the microbuckets, as a result of longitudinal amplitude dependent tune spread of the electron beam. Here we present a remarkable feature of phase mixing or filamentation in RF or optical buckets. As we will see soon, given an initial DC mono-energetic beam, there will be an equilibrium phase space distribution after phase mixing in the bucket. We find that in this final steady state, the beam current distribution has little dependence on the bucket height. This feature is favorable for the SSMB PoP II, as the requirement on the modulation laser power can then be much relaxed compared to PoP I. This effect is also of relevance to the injection process of the final real SSMB storage ring.

As the phase mixing is a rather fast process compared to radiation damping, we consider only the symplectic dynamics in this section for simplicity. The symplectic longitudinal dynamics of a particle in a storage ring with a single RF system, in SSMB a laser modulator, can be modeled by the well-known "standard map" [16]

$$
\begin{cases}
I_{n+1} = I_n + K \sin \theta_n, \\
\theta_{n+1} = \theta_n + I_{n+1},
\end{cases}
\tag{5.15}
$$

in which

$$
\theta = k_{\mathrm{RF}}z, \quad I = R_{56}k_{\mathrm{RF}}\delta, \quad K = \frac{V_{\mathrm{RF}}}{E_0} R_{56}k_{\mathrm{RF}},
\tag{5.16}
$$

with $R_{56} = -\eta C_0$. Note that K in this section is not the undulator parameter. Equation (5.15) can be described with the pendulum Hamiltonian driven by a periodic perturbation

$$H(I, \theta, t) = \frac{1}{2} I^2 + K \cos\theta \sum_{n=-\infty}^{\infty} \cos(2\pi nt). \qquad (5.17)$$

The dynamics is given by a sequence of free propagations interleaved with periodic kicks. For $K \neq 0$, the dynamics is non-integrable and chaotic. But for a K much smaller than 1, which is the case for usual storage rings working in the longitudinal weak focusing regime, the motion is close to integrable and the differences in Eq. (5.15) can be approximately replaced by differentiation, and the Hamiltonian Eq. (5.17) can be replaced by a pendulum Hamiltonain

$$H = \frac{1}{2} I^2 + K \cos\theta. \qquad (5.18)$$

The separatrix of the pendulum bucket is $H = K$ with a bucket half-height of $2\sqrt{K}$. Or in unit of δ, the bucket half-height is

$$\delta_{\frac{1}{2}} = \frac{2\sqrt{V_{RF} R_{56} k_{RF}}}{|R_{56} k_{RF}|} \approx \frac{1}{\beta_{zS} k_{RF}}, \qquad (5.19)$$

where β_{zS} is the longitudinal beta function at the RF as analyzed in Sect. 2.1.2. The synchrotron tune is

$$\nu_s \approx -\text{sgn}(K) \frac{\sqrt{K}}{2\pi}. \qquad (5.20)$$

When K is large, the strongly chaotic dynamics can also be used for interesting applications, for example applying the bucket purification to generate short bunches as proposed in Ref. [17].

Figure 5.20 shows a simulation result of the evolution in the longitudinal phase space of a mono-energetic DC beam after injection into RF or optical buckets described by Eq. (5.15). We have chosen to observe the beam in the middle of the RF kick so the beam distribution in the longitudinal phase space is upright. As can be seen, there is a steady-state beam distribution due to phase mixing in the bucket. Note that the bucket center is at $\theta = \pi$ when $K > 0$. If $K < 0$, then the bucket center will be at $\theta = 0$.

As the longitudinal form factor, thus the coherent radiation power, depends more directly on the beam current (namely the longitudinal coordinate z of the electrons) rather than the energy spread, now we try to get an analytical formula for the steady-state beam current. For convenience, we shift the bucket center to the origin, which means $\theta - \pi \rightarrow \theta$, or a sign change of K. What we want to know is the steady-state distribution of θ, i.e., $f(\theta, t \rightarrow \infty)$. In action-angle (ϕ, J) phase space, the distribution function evolves according to

$$f(\phi, J, t) = f(\phi - \omega(J)t, J, 0). \qquad (5.21)$$

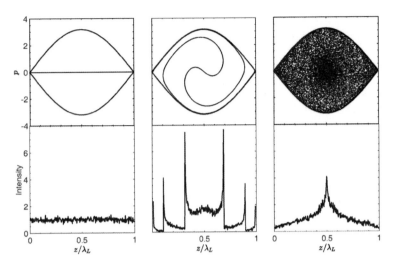

Fig. 5.20 Phase mixing (filamentation or decoherence) of a mono-energetic particle beam trapped in RF or optical bucket with $K = 0.01$. Up: particle distribution in longitudinal phase space, with red curves being the separatrices. Bottom: the corresponding beam current distribution

Fig. 5.21 A plot to help better understand Eq. (5.23)

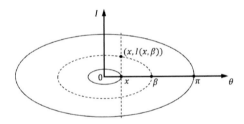

When there is a tune dependence $\omega(J)$ on J, then in the limit of $t \to \infty$, the steady-state distribution depends only on the initial distribution of action J as a result of phase mixing,

$$f(\phi, J, t \to \infty) = \frac{1}{2\pi} \int_0^{2\pi} f(\phi, J, t = 0)d\phi. \tag{5.22}$$

The final angle for each action J will uniformly distributed in $[0, 2\pi)$. As shown in Fig 5.21, after reaching the steady state, the percentage of the particles with $\theta \leq x$ for $0 < x < \pi$ is

$$P(\theta \leq x) = \frac{x}{\pi} + \int_x^{\pi} \left(1 - \frac{\phi(x, I(x, \beta))}{\pi} \right) \frac{1}{\pi} d\beta, \tag{5.23}$$

in which $I(x, \beta)$ represents the I-coordinate of a point on the (ϕ, J) phase space trajectory traversing $(\beta, 0)$ with a θ-coordinate of x.

After getting $P(\theta \le x)$, the current distribution can then be calculated according to

$$f(\theta) = \left. \frac{\partial P}{\partial x} \right|_{x=\theta}. \tag{5.24}$$

However, $\phi(x, I(x, \beta))$ has a complex form, and it is hard to get a simple analytical expression for $f(\theta)$. Here we simplify the discussion by approximating all the phase space trajectories in the bucket by ellipses to arrive at an analytical formula for $f(\theta)$. For an ellipse phase space trajectory, we have

$$\phi(x, I(x, \beta)) = \arccos \frac{x}{\beta}. \tag{5.25}$$

Note that the result in Eq. (5.25) has no dependence on K. For a real RF or optical bucket, there is a dependence of $\phi(x, I(x, \beta))$ on K, but the dependence is weak, especially for trajectories close to the bucket center. So we expect our approximated Eq. (5.25) is valid to a large extent. Substituting Eq. (5.25) into Eqs. (5.23) and (5.24), we have

$$f(\theta) = \left. \frac{\partial P}{\partial x} \right|_{x=\theta} = \int_{\theta}^{\pi} \left(\frac{1}{\pi \sqrt{1 - \left(\frac{\theta}{\beta} \right)^2}} \frac{1}{\beta} \right) \frac{1}{\pi} d\beta = \frac{1}{\pi^2} \ln \left| \frac{\pi + \sqrt{\pi^2 - \theta^2}}{\theta} \right|. \tag{5.26}$$

Note that our simplified theoretically current distribution $f(\theta)$ is independent of K, which means the steady-state current distribution is independent of the bucket height.

Figure 5.22 shows the simulation result of the steady-state current distribution under different K, i.e., different bucket heights, and simultaneously our simplified theoretical distribution Eq. (5.26). As can be seen that indeed the steady-state current distribution has little dependence on the bucket height, and our simplified analysis is quite accurate. Note that the origin is not shifted in the plot.

The analysis reveals a remarkable feature of phase mixing in RF or optical bucket, i.e., the final steady-state current distribution after a mono-energetic beam getting

Fig. 5.22 The steady-state current distribution after phase mixing in RF or optical bucket, with different K. In each simulation, 2×10^5 particles have been tracked for 2×10^5 turns. Also presented in the figure is the theoretical prediction given by Eq. (5.26)

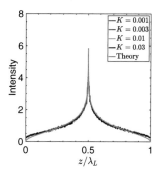

Fig. 5.23 The steady-state current distribution after phase mixing in RF or optical bucket, with an increase of K from 0.001 to 0.03 in two consecutive steps. In each step, 2×10^5 particles have been tracked for 2×10^5 turns

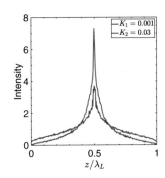

trapped by RF or optical buckets has little dependence on the bucket height. This is helpful for our Quasi-SSMB experiment since it means the requirement on the modulation laser power is not that demanding. A bucket height several times of the natural energy spread is sufficient.

The above result is based on a constant RF voltage in the phase mixing process, it can be anticipated that more particles will be bunched closer to the bucket center phase when we increase K after the beam reach its steady-state distribution after phase mixing. Similar steps to the above section can be invoked for calculating the new steady-state current distribution. A transformation of the action when K changes is all that needed. Figure 5.23 shows the simulation result of the steady-state current distribution by increasing K in two consecutive steps from 0.001 to 0.03. As can be seen, the current are more concentrated to the center after the increase of K.

A discrete change of K can boost bunching as shown in Fig. 5.23. However, it is not without sacrifice, as the filamentation process will result in longitudinal emittance growth. This emittance increase is unwanted in some cases. As well-studied in RF gymnastics [18], an adiabatic change of RF voltage or lattice parameters can manipulate the bunch length while preserving the longitudinal emittance. Similar ideas can also be applied to boost microbunching with little emittance growth. A simulation of trapping of microbunch with K linearly ramped from 1×10^{-6} to 1×10^{-2} is shown in Fig. 5.24. Note the drastic difference between Figs. 5.24 and 5.20. The spirit of adiabatic buncher [19, 20] is the same with adiabatic trapping introduced here, for enhancing microbunching while preserving longitudinal emittance which is useful for FEL and inverse FEL. The adiabatic trapping mechanism can also be applied in the beam injection of the SSMB or other storage rings whose momentum aperture is of concern. It is interesting to note the connection of adiabatic trapping with the microbunching process in a high-gain FEL [21–23].

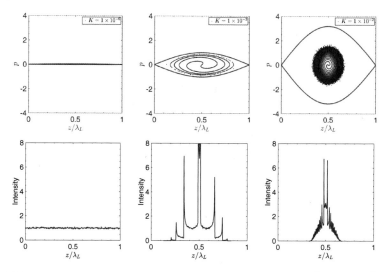

Fig. 5.24 Trapping of particles with a linear increase of K from 1×10^{-6} to 1×10^{-2} in 10^4 turns. Up: particle distribution in longitudinal phase space, with red curves being the separatrices. Bottom: the corresponding beam current distribution

5.3.2 Experimental Parameters Choice

As the quantum diffusion of z is large for the MLS lattice (26 nm RMS per turn at 250 MeV corresponding to optics in Fig. 5.2), it is not feasible to realize true SSMB inside a 1064 nm wavelength microbucket at the MLS. Therefore, the goal of SSMB PoP II is to accomplish microbunching for 100 to 1000 consecutive turns to reach a quasi steady state. Based on the beam physics and noises analysis, the tentative experimental parameters choice is as shown in Table 5.2. We have conducted numerical simulations based on the parameters set, from which we observe that the typical evolution of electrons in PoP II experiment can be divided into several stages.

- I: with the modulation laser turned on, the bunching factor reaches the maximum after about one quarter of the synchrotron oscillation period;
- II: after several synchrotron oscillation periods, the whole microbuckets are filled with particles like that shown in the right part of Fig. 5.20 as a result of phase mixing. Bunching factor after reaching this quasi-steady state will be stable if there is no quantum excitation or other diffusion effects;
- III: due to quantum excitation and various diffusion effects, energy spread starts to increase and particles continue to leak out the microbuckets and begin to hit on the vacuum pipe and become lost. Bunching factor in this stage decreases;
- IV: after a while, all the particles are lost in the end.

To make the experiment more realistic, each time we fire the laser, we only want to accomplish the above stages I and II, but avoid III and IV, i.e., to avoid particle loss, otherwise it will be too time-consuming to do the experiment. This is based on

Table 5.2 Tentative parameters of the Quasi-SSMB experiment to be conducted at the MLS

Parameter	Value	Description
E_0	250 MeV	Beam energy
C_0	48 m	Ring circumference
f_{RF}	500 MHz	RF frequency
η	$\lvert\eta\rvert \leq 2 \times 10^{-5}$	Phase slippage factor
τ_δ	180 ms@250 MeV	Longitudinal radiation damping time
σ_δ	1.8×10^{-4}@250 MeV	Natural energy spread
ϵ_x	31 nm@250 MeV	Horizontal emittance
λ_u	125 mm	Undulator period length
N_u	32	Number of undulator periods
L_u	4 m	Undulator length
K	2.5	Undulator parameter
λ_L	1064 nm	Modulation laser wavelength
Z_R	$\sim \frac{L_u}{3}$	Rayleigh length
P_{peak}	≥ 10 kW	Modulation laser peak power
$\delta_{\frac{1}{2}}$	$\geq 1.5\sigma_\delta$	Microbucket half-height
λ_R	1064 nm	Radiation wavelength
b_1	≥ 0.01	Bunching factor (1064 nm) in quasi-steady state

the fact that preparing the beam and storage ring state is time-consuming, while the particle can be lost in milli seconds with the laser keep firing. This is why we aim for preserving mircorbunching for 10^3 turns, instead of 10^6 turns or a longer time. The experiment is under preparation and more progress will be reported in the future. One thing worth mentioning is that the second-harmonic bunching in the quasi-steady state is negligible at the MLS, therefore, fundamental frequency radiation detection is needed in SSMB PoP II.

References

1. Deng X, Chao A, Feikes J, Hoehl A, Huang W, Klein R, Kruschinski A, Li J, Matveenko A, Petenev Y et al (2021) Experimental demonstration of the mechanism of steady-state microbunching. Nature 590(7847):576–579
2. Kruschinski A, Feikes J, Li J, Ries M, Deng X, Hoehl A, Klein R (2023) Exploring the necessary conditions for steady-state microbunching at the metrology light source. In: Proceedings of the 14th international particle accelerator conference (IPAC'23), Vinece, Italy, 2023. JACoW, Geneva, p MOPA176
3. Tang C, Deng X, Chao A, Huang W, Rui T, Feikes J, Li J, Ries M, Hoehl A, Ratner D et al (2018) An overview of the progress on ssmb. In: Proceedings of the 60th ICFA advanced beam dynamics workshop on future light sources (FLS'18), Shanghai, China, 2018. JACoW, Geneva, pp 166–170

4. Klein R, Brandt G, Fliegauf R, Hoehl A, Müller R, Thornagel R, Ulm G, Abo-Bakr M, Feikes J, Hartrott MV, Holldack K, Wüstefeld G (2008) Operation of the metrology light source as a primary radiation source standard. Phys Rev ST Accel Beams 11:110701

5. Feikes J, von Hartrott M, Ries M, Schmid P, Wüstefeld G, Hoehl A, Klein R, Müller R, Ulm G (2011) Metrology light source: the first electron storage ring optimized for generating coherent thz radiation. Phys Rev ST Accel Beams 14:030705

6. Ries M (2014) Nonlinear momentum compaction and coherent synchrotron radiation at the metrology light source. Dissertation, Humboldt University of Berlin. PhD thesis, Humboldt-Universität zu Berlin, Berlin

7. Piwinski A (1974) Intra-beam-scattering. In: Proceedings of the 9th international conference on high energy accelerators, Stanford, CA, USA, p 405

8. Bjorken JD, Mtingwa SK (1982) Intrabeam scattering. Part Accel 13(FERMILAB-PUB-82-47-THY):115–143

9. Bernardini C, Corazza G, Di Giugno G, Ghigo G, Haissinski J, Marin P, Querzoli R, Touschek B (1963) Lifetime and beam size in a storage ring. Phys Rev Lett 10(9):407

10. Chao AW (1993) Physics of collective beam instabilities in high energy accelerators. Wiley

11. Chao A (2022) Focused laser, unpublished note

12. Borland M (2000) Elegant: a flexible sdds-compliant code for accelerator simulation, technical report, Argonne National Lab

13. Tanaka T, Kitamura H (2001) Spectra: a synchrotron radiation calculation code. J Synchrotron Radiat 8(6):1221–1228

14. Falkenstern F, Hoffmann F, Kuske P, Kuszynski J (2009) Bunch view-a fast and accurate bunch-by-bunch current monitor. Proceedings of DIPAC09, Basel, Switzerland, pp 1–3

15. Kruschinski A (2021) A fundamental mode detection scheme for the steady-state microbunching proof-of-principle experiment at the metrology light source

16. Chirikov BV (1979) A universal instability of many-dimensional oscillator systems. Phys Rep 52(5):263–379

17. Jiao Y, Ratner DF, Chao AW (2011) Terahertz coherent radiation from steady-state microbunching in storage rings with x-band radio-frequency system. Phys Rev ST Accel Beams 14:110702

18. Garoby R (2011) Rf gymnastics in synchrotrons. arXiv preprint arXiv:1112.3232

19. Hemsing E, Xiang D (2013) Cascaded modulator-chicane modules for optical manipulation of relativistic electron beams. Phys Rev ST Accel Beams 16:010706

20. Sudar N, Musumeci P, Gadjev I, Sakai Y, Fabbri S, Polyanskiy M, Pogorelsky I, Fedurin M, Swinson C, Kusche K, Babzien M, Palmer M (2018) Demonstration of cascaded modulator-chicane microbunching of a relativistic electron beam. Phys Rev Lett 120:114802

21. Kroll NM, McMullin WA (1978) Stimulated emission from relativistic electrons passing through a spatially periodic transverse magnetic field. Phys Rev A 17:300–308

22. Kondratenko A, Saldin E (1980) Generating of coherent radiation by a relativistic electron beam in an ondulator. Part Accel 10:207–216

23. Bonifacio R, Pellegrini C, Narducci L (1984) Collective instabilities and high-gain regime free electron laser. In: AIP conference proceedings, vol 118. American Institute of Physics, pp 236–259

Chapter 6
Summary

In this final chapter, we give a brief summary of the dissertation, and present some useful results for practitioners.

6.1 Summary of the Dissertation

The contribution of this dissertation consists of three parts: in Chaps. 2 and 3, we answer the question of how to realize SSMB; in Chap. 4, we investigate what radiation characteristics can be obtained from the formed SSMB; and in Chap. 5, we experimentally demonstrate the working mechanism of SSMB in a real machine for the first time.

In Chap. 2, to account for the impact of local phase slippage factors on beam dynamics in a quasi-isochronous electron storage ring, we have developed and applied the Courant-Snyder formalism in longitudinal dimension to derive new formulae of bunch length, energy spread and longitudinal emittance beyond the classical scaling laws. The method of optimizing the global and local phase slippages simultaneously to minimize the longitudinal β function at the bending magnets has been proposed based on the analysis, to generate an ultrashort bunch length and ultrasmall longitudinal emittance, as required by SSMB. Further, we have derived the scaling law of the theoretical minimum bunch length and longitudinal emittance with respect to the bending radius and angle of the bending magnet. The use of transverse gradient bends for minimizing the longitudinal emittance has also been investigated. The application of multiple RF cavities, or laser modulators in an SSMB storage ring, for longitudinal strong focusing has been discussed using the same formalism with important observations made. Considering the momentum compaction of a laser modulator, its thick-lens linear and nonlinear maps have been derived and simulated for a more accurate modeling of beam dynamics in it. We have also studied the

X. Deng, *Theoretical and Experimental Studies on Steady-State Microbunching*, Springer Theses, https://doi.org/10.1007/978-981-99-5800-9_6

application of the higher-order terms of phase slippage for high-harmonic bunching and longitudinal dynamic aperture optimization. Based on the investigations in this chapter, we have presented in Table 6.1 an example parameters set of a longitudinal weak focusing SSMB storage ring for high-power infrared radiation generation.

In Chap. 3, we have presented a concise analysis of the bending magnet-induced passive bunch lengthening from transverse emittance of the particle beam. After that, we have generalized the analysis and proved three theorems on the active applications of transverse-longitudinal coupling (TLC) for efficient harmonic generation or bunch length compression. These theorems dictate the relation between the modulation lick strength and the lattice optical functions at the modulator and radiator, respectively. Further, we have analyzed the contribution of modulators to the vertical emittance from quantum excitation, to obtain a self-consistent evaluation of the required modulation laser power in applying these TLC schemes in a storage ring. These theorems and related analysis provide the theoretical basis for the application of TLC in SSMB to lower the requirement on the modulation laser power, by taking advantage of the fact that the vertical emittance in a planar ring is rather small. The relation between our TLC analysis and the transverse-longitudinal emittance exchange is also briefly discussed. In addition to the investigation on linear TLC dynamics, we have also reported the first experimental validation of particle energy widening and distortion by a nonlinear TLC effect in a quasi-isochronous ring, which originates from an average path-length dependence on the betatron oscillation amplitudes. The result could be important for quasi-isochronous rings, SSMB, nonscaling fixed-field alternate gradient accelerators, etc., where very small phase slippage factor or large chromaticity is required. Based on the investigations in this chapter, we have presented in Table 6.2 an example parameters set of a transverse-longitudinal coupling SSMB storage ring for high-power EUV and soft X-ray radiation generation.

In Chap. 4, we have presented theoretical and numerical studies of the average and statistical property of the coherent radiation from SSMB. Our results show that kW-level average power of 13.5 nm-wavelength EUV radiation can be obtained from an SSMB ring, provided that an average current of 1 A and bunch length of 3 nm microbunch train can be formed at the radiator. Such a high-power EUV source is a promising candidate to fulfill the urgent need of semiconductor industry for EUV lithography. Together with its narrowband feature, the EUV photon flux can reach $10^{15} \sim 10^{16}$ phs/s within a 0.1 meV energy bandwidth, which is appealing for fundamental condensed matter physics research. In the theoretical investigation, we have generalized the definition and derivation of the transverse form factor of an electron beam which can quantify the impact of its transverse size on the coherent radiation. In particular, we have shown that the narrowband feature of SSMB radiation is strongly correlated with the finite transverse electron beam size. Considering the pointlike nature of electrons and quantum nature of radiation, the coherent radiation fluctuates from microbunch to microbunch, or for a single microbunch from turn to turn. Some important results concerning the statistical property of SSMB radiation have been presented, with a brief discussion on its potential applications for example the beam diagnostics. The presented work is of value for the development of SSMB and better serve the potential synchrotron radiation users. In addition, it also sheds light on understanding the radiation characteristics of free-electron lasers, coherent harmonic generation, etc.

In Chap. 5, we have reported the first demonstration of the mechanism of SSMB at the Metrology Light Source in Berlin. We have shown that electron bunches stored in a quasi-isochronous ring can yield sub-micrometre microbunching and narrow-band coherent radiation, one complete revolution after energy modulation induced by a 1064 nm wavelength laser, and this microbunching can preserve for multiple turns. These results verify that the optical phases, i.e, the longitudinal coordinates, of electrons can be correlated turn by turn in a storage ring at a precision of sub-laser wavelengths. On the basis of this phase correlation, we expect that SSMB will be realized by applying a phase-locked laser that interacts with the electrons turn by turn. This demonstration represents the first milestone towards the implementation of an SSMB-based high-power, high-repetition photon source.

6.2   Useful Formulas and Example Parameters for SSMB Storage Rings

To make our investigations more useful for practitioners, especially concerning the parameters choice for an SSMB storage ring, here we present some important formulas. Generally we group our formulas into two categories, i.e., a longitudinal weak focusing storage ring for a desired radiation wavelength $\lambda_R \gtrsim 100$ nm, and a transverse-longitudinal coupling storage ring for a desired radiation wavelength 1 nm $\lesssim \lambda_R \lesssim 100$ nm. In each category, we have presented an example parameters set for the corresponding SSMB storage ring.

6.2.1   Longitudinal Weak Focusing SSMB

The relation of bending radius ρ and magnetic flux density B of the bending magnet is

$$\frac{1}{\rho} = 0.2998 \frac{B[\mathrm{T}]}{E_0[\mathrm{GeV}]}, \tag{6.1}$$

with E_0 the electron energy.

Assuming that the storage ring consists of isomagnets, then the radiation loss of an electron per turn is

$$U_0 = C_\gamma \frac{E_0^4}{\rho_{\mathrm{ring}}}, \tag{6.2}$$

with $C_\gamma = 8.85 \times 10^{-5} \frac{\mathrm{m}}{\mathrm{GeV}^3}$, ρ_{ring} the bending radius of bending magnets in the ring.

The horizontal, vertical and longitudinal radiation damping constants for a planar uncoupled ring are

$$\alpha_H = \frac{U_0}{2E_0}(1 - \mathcal{D}),$$

$$\alpha_V = \frac{U_0}{2E_0}, \tag{6.3}$$

$$\alpha_L = \frac{U_0}{2E_0}(2 + \mathcal{D}),$$

where $\mathcal{D} = \dfrac{\oint \frac{(1-2n)D_x}{\rho^3} ds}{\oint \frac{1}{\rho^2} ds}$, with $n = -\frac{\rho}{B}\frac{\partial B}{\partial \rho}$ the field gradient index and D_x is the horizontal dispersion. Nominally for a planar uncoupled ring using bending magnets with no transverse gradient, we have $\mathcal{D} \ll 1$.

The horizontal, vertical and longitudinal radiation damping times are

$$\tau_{H,V,L} = \frac{C_0/c}{\alpha_{H,V,L}}, \tag{6.4}$$

with C_0 the ring circumference and c the speed of light in vacuum.

The natural energy spread of electron beam in a longitudinal weak focusing ring is

$$\sigma_{\delta S} = \sqrt{\frac{C_q}{J_s}\frac{\gamma^2}{\rho}}, \tag{6.5}$$

with $C_q = \frac{55\lambda_e}{32\sqrt{3}} = 3.8319 \times 10^{-13}$ m, $\lambda_e = \frac{\lambda_e}{2\pi} = 386$ fm is the reduced Compton wavelength of electron, $J_s = 2 + \mathcal{D}$ is the longitudinal damping partition number, γ is the Lorentz factor.

The natural bunch length at the laser modulator is

$$\sigma_{zS} = \sigma_{\delta S}\beta_{zS}, \tag{6.6}$$

where β_{zS} is the longitudinal beta function at the laser modulator to be given soon.

The effective modulation voltage of a laser modulator using a planar undulator is [1]

$$V_L = \frac{[JJ]K}{\gamma}\sqrt{\frac{4P_L Z_0 Z_R}{\lambda_L}}\tan^{-1}\left(\frac{L_u}{2Z_R}\right). \tag{6.7}$$

in which $[JJ] = J_0(\chi) - J_1(\chi)$ and $\chi = \frac{K^2}{4+2K^2}$, J_n is the n-th order Bessel function of the first kind, $K = \frac{eB_0}{m_e c k_u} = 0.934 \cdot B_0[\text{T}] \cdot \lambda_u[\text{cm}]$ is the undulator parameter, determined by the peak magnetic flux density B_0 and period λ_u of the undulator, P_L is the modulation laser power, $Z_0 = 376.73$ Ω is the impedance of free space, Z_R is the Rayleigh length of the laser, L_u is the undulator length.

The linear energy chirp strength around zero-crossing phase is related to the laser and modulator undulator parameters according to

$$h = \frac{eV_L}{E_0}k_L = \frac{e[JJ]K}{\gamma^2 mc^2}\sqrt{\frac{4P_L Z_0 Z_R}{\lambda_L}}\tan^{-1}\left(\frac{L_u}{2Z_R}\right)k_L,\tag{6.8}$$

where $k_L = 2\pi/\lambda_L$ is the wavenumber of the modulation laser.

Linear stability of the longitudinal motion requires

$$0 < h\eta C_0 < 4,\tag{6.9}$$

where η is the phase slippage factor of the ring.

Considering the fact that the modulation waveform is sinusoidal and the longitudinal dynamics is more accurately modeled by a "standard kick map", to avoid strong chaotic dynamics, an empirical criterion is

$$0 < h\eta C_0 \lesssim 0.1.\tag{6.10}$$

In a longitudinal weak focusing ring ($\nu_s \ll 1$), the synchrotron tune is

$$\nu_s \approx \frac{\eta}{|\eta|}\frac{\sqrt{h\eta C_0}}{2\pi}.\tag{6.11}$$

In a longitudinal weak focusing ring, the longitudinal beta function at the laser modulator is

$$\beta_{zS} \approx \sqrt{\frac{\eta C_0}{h}}.\tag{6.12}$$

The micro-bucket half-height is

$$\hat{\delta}_{\frac{1}{2}} = \frac{2}{\beta_{zS}k_L}.\tag{6.13}$$

If there is a single RF or laser modulator in the ring, and $J_s = 2$, then the theoretical minimum bunch length and longitudinal emittance in a longitudinal weak focusing ring with respect to the bending radius ρ and angle θ of each bending magnet are

$$\sigma_{z,\min}[\mu m] \approx 4.93\rho^{\frac{1}{2}}[m]E_0[GeV]\theta^3[rad],$$
$$\epsilon_{z,\min}[nm] \approx 8.44E_0^2[GeV]\theta^3[rad].\tag{6.14}$$

Scaling law of the horizontal emittance in an SSMB storage ring is

$$\epsilon_x[nm] \approx -366.5E_0^2[GeV]\theta^3[rad]\left[\frac{1}{9}\tan\left(\frac{\Phi_x}{2}\right) + \frac{1}{10}\cot\left(\frac{\Phi_x}{2}\right)\right],\tag{6.15}$$

with Φ_x the horizontal betatron phase advance per cell which usually lies in $(\pi, 2\pi)$. The above scaling is derived by considering only the contribution of main cells, and ignoring that from the matching section.

Coherent undulator radiation power at the odd-H-th harmonic from a transversely-round electron beam is

$$P_{H,\text{peak}}[\text{kW}] = 1.183 N_u H \chi [JJ]_H^2 FF_\perp(S)|b_{z,H}|^2 I_P^2[\text{A}], \qquad (6.16)$$

where N_u is the number of undulator periods, $[JJ]_H^2 = \left[J_{\frac{H-1}{2}}(H\chi) - J_{\frac{H+1}{2}}(H\chi) \right]^2$, with $\chi = \frac{K^2}{4+2K^2}$, and the transverse form factor is

$$FF_\perp(S) = \frac{2}{\pi} \left[\tan^{-1} \left(\frac{1}{2S} \right) + S \ln \left(\frac{(2S)^2}{(2S)^2 + 1} \right) \right], \qquad (6.17)$$

with $S = \frac{\sigma_\perp^2 \frac{\omega}{c}}{L_u}$ and $\sigma_\perp$ the RMS transverse electron beam size, $b_{z,H}$ is the bunching factor at the H-th harmonic, and I_P is the peak current.

The relative fluctuation of coherent radiation power considering the pointlike nature of electrons is

$$\frac{\text{Var}\left[|b(\mathbf{k})|^2\right]}{\langle|b(\mathbf{k})|^2\rangle^2} = \frac{2}{N_e} \left(\frac{|\bar{b}(\mathbf{k})|^2 + \text{Re}\left[\bar{b}(2\mathbf{k})\bar{b}^2(-\mathbf{k})\right]}{|\bar{b}(\mathbf{k})|^4} - 2 \right) + O\left(\frac{1}{N_e^2}\right),$$

$$(6.18)$$

where $b(\mathbf{k})$ is the bunching factor at the wavevector $\mathbf{k}$, and N_e is the number of electrons.

Based on the above formulas, here we present an example parameters set in Table 6.1 of a longitudinal weak focusing SSMB storage ring, aimed for high-power infrared radiation generation. As can be seen, such a compact SSMB storage ring can be used for power amplification of the injected seed laser. The requirement on the stored laser power is easy to realize in practice. All the other parameters are also within practical range. A sharp reader may notice that the microbucket half-height is only twice the natural energy spread of the electron beam. Therefore, in addition to these shallow microbuckets, we need a larger bucket, for example a barrier bucket formed by an induction linac, to constrain the particles in the ring to ensure a large enough beam lifetime.

6.2.2 Transverse-Longitudinal Coupling SSMB

For a transverse-longitudinal coupling (TLC) based SSMB, or a generalized longitudinal strong focusing SSMB [2], using TEM00 mode laser modulator for energy modulation, we have the following important formulas.

Table 6.1 Example parameters set of a longitudinal weak focusing SSMB storage ring for infrared radiation generation

Parameter	Value	Description
E_0	250 MeV	Beam energy
C_0	50 m	Circumeference
η	4×10^{-6}	Phase slippage factor
ρ_{ring}	0.6 m	Bending radius of dipoles in the ring
B_{ring}	1.39 T	Bending field in the ring
θ	$\frac{\pi}{7}$	Bending angle of each dipole
$\sigma_{\delta S}$	2.76×10^{-4}	Natural energy spread
$\sigma_{z,\text{lim}}$	86 nm	Theoretical lower bunch length limit
λ_L	1064 nm	Modulation laser wavelength
h	500 m^{-1}	Energy chirp strength
σ_{zS}	175 nm	Natural bunch length
$\delta_{\frac{1}{2}}$	5.36×10^{-4}	Microbuket half-height
$\lambda_{u\text{Mod}}$	5 cm	Modulator undulator period
$B_{0\text{Mod}}$	0.92 T	Modulator peak magnetic field
$L_{u\text{Mod}}$	1 m	Modulator undulator length
$P_L(Z_R = \frac{L_u}{3})$	24 kW	Modulation laser power
g	5×10^3	Optical enhancement cavity gain
P_{in}	4.8 W	Injection laser power
$\lambda_R = \lambda_L$	1064 nm	Radiation wavelength
b_1	0.59	Bunching factor
$\sigma_\perp$	100 μm	Transverse electron beam size at the radiator
$\lambda_{u\text{Rad}}$	5 cm	Radiator undulator period
$B_{0\text{Rad}}$	0.92 T	Radiator peak magnetic field
$L_{u\text{Rad}}$	2 m	Radiator length
P_R	1 kW @ $I_P = 0.55$ A	Radiation peak/average power

Relation between energy chirp strength and optical functions at the modulator and radiator

$$h^2(\text{Mod})\mathcal{H}_y(\text{Mod})\mathcal{H}_y(\text{Rad}) \geq 1, \qquad (6.19)$$

where $\mathcal{H}_y$ is a chromatic function quantifying the contribution of vertical emittance to bunch length.

Put the above relation in another way,

$$h \geq \frac{\epsilon_y}{\sigma_{zy}(\text{Mod})\sigma_{zy}(\text{Rad})}. \tag{6.20}$$

Bunching factor at the n-th laser harmonic in TLC SSMB at the radiator

$$b_n = \left(\sum_{m=-\infty}^{\infty} J_m(n) \exp\left[-((n-m)k_L\sigma_z(\text{Mod}))^2/2 \right] \right) \exp\left[-(nk_L\sigma_z(\text{Rad}))^2/2 \right], \tag{6.21}$$

where $\sigma_z(\text{Mod}) = \sqrt{\epsilon_z\beta_z(\text{Mod}) + \epsilon_y\mathcal{H}_y(\text{Mod})}$ and $\sigma_z(\text{Rad}) = \sqrt{\epsilon_y\mathcal{H}_y(\text{Rad})}$ are the linear bunch length at the modulator and radiator, respectively.

Contribution of two modulators to ϵ_y from quantum excitation

$$\Delta\epsilon_y(\text{Mod}) = 2 \times \frac{55}{96\sqrt{3}} \frac{\alpha_F \lambda_e^2 \gamma^5}{\alpha_V} \frac{\mathcal{H}_y(\text{Mod})}{\rho_{0\text{Mod}}^3} \frac{4}{3\pi} L_u, \tag{6.22}$$

where $\alpha_F = \frac{1}{137}$ is the fine-structure constant.

Assuming $\epsilon_y = \Delta\epsilon_y(\text{Mod})$, which means the vertical emittance is solely from the two modulators, then the required modulation laser power and modulator length scaling are

$$P_L[\text{kW}] \approx 5.67 \frac{\lambda_L^{\frac{7}{3}}[\text{nm}]E_0^{\frac{8}{3}}[\text{GeV}]B_{0\text{Mod}}^{\frac{7}{3}}[\text{T}]}{\sigma_z^2(\text{Rad})[\text{nm}]B_{\text{ring}}[\text{T}]},$$

$$L_u[\text{m}] \approx 57 \frac{B_{\text{ring}}[\text{T}]\epsilon_y[\text{pm}]}{\mathcal{H}_y(\text{Mod})[\mu\text{m}]B_{0\text{Mod}}^3[\text{T}]}, \tag{6.23}$$

where $B_{0\text{Mod}}$ is the peak magnetic flux density of the modulator undulator, B_{ring} is the magnetic flux density of bending magnets in the ring. The above scaling laws are accurate when $K_u > \sqrt{2}$. For the more general case, refer to Eq. (3.56).

Based on the presented formulas, here we present an example parameters set in Table 6.2 of a TLC SSMB storage ring, aimed for high-power EUV and soft X-ray radiation. It can be seen that as long as we can realize a coasting beam of 1.5 A average current, and an optical cavity stored power of $\gtrsim 100$ kW, we can realize 1 kW average power 13.5 nm EUV and 6.75 nm soft X-ray radiation. All the other parameters applied should be realizable, including the small ϵ_y considering IBS. Even if we can only realize an average beam current of 1 A or less, we can take advantage of the fact that $P_{\text{coh}} \propto I_P^2$ to realize an average radiation power of kW level, by decreasing the filling factor of electron beam in the ring but increasing the peak current as long as the value is below the collective instability threshold.[1] Since there is no requirement on the longitudinal emittance for a coasting beam, thus no requirement on the fine control of longitudinal β function, the circumference of this ring has great flexibility, which means the ring can be very compact, for example a

---

[1] Private communication with Alex Chao.

Table 6.2 Example parameters set of a transverse-longitudinal coupling SSMB storage ring for EUV and soft X-ray radiation generation

	Parameter	Value	Description
	E_0	800 MeV	Beam energy
	C_0	~100 m	Circumference
	B_{ring}	1.33 T	Bending field in the ring
	ρ_{ring}	2 m	Bending radius in the ring
	$\sigma_{\delta S}$	4.85×10^{-4}	Natural energy spread
	ϵ_y	2 pm	Vertical emittance
	λ_L	270 nm	Modulation laser wavelength
	$\sigma_\perp$	10 μm	Transverse electron beam size at the radiator
EUV (13.5 nm)	σ_z (Rad)	2 nm	Linear bunch length at the radiator
	σ_{zy} (Mod)	1.85 μm	Bunch lengthening from ϵ_y at the modulator
	h	541 m^{-1}	Energy chirp strength
	$\lambda_{u\text{Mod}}$	0.5 m	Modulator undulator period
	$B_{0\text{Mod}}$	0.039 T	Modulator peak magnetic flux density
	$L_{u\text{Mod}}$	1.5 m	Modulator length
	$P_L(Z_R = \frac{L_u}{3})$	141 kW	Modulation laser power
	$\lambda_R = \frac{\lambda_L}{20}$	13.5 nm	Radiation wavelength
	b_{20}	0.11	Bunching factor
	$\lambda_{u\text{Rad}}$	2 cm	Radiator undulator period
	$B_{0\text{Rad}}$	1.15 T	Radiator peak magnetic flux density
	$L_{u\text{Rad}}$	3.2 m	Radiator length
	P_R	1 kW @ $I_P = 1.5$ A	Radiation peak/average power
Soft X-ray (6.75 nm)	σ_z (Rad)	1 nm	Linear bunch length at the radiator
	σ_{zy} (Mod)	1.85 μm	Bunch lengthening from ϵ_y at the modulator
	h	1082 m^{-1}	Energy chirp strength
	$\lambda_{u\text{Mod}}$	0.5 m	Modulator undulator period
	$B_{0\text{Mod}}$	0.039 T	Modulator peak magnetic flux density
	$L_{u\text{Mod}}$	1.5 m	Modulator length
	$P_L(Z_R = \frac{L_u}{3})$	564 kW	Modulation laser power
	$\lambda_R = \frac{\lambda_L}{40}$	6.75 nm	Radiation wavelength
	b_{40}	0.085	Bunching factor
	$\lambda_{u\text{Rad}}$	1.5 cm	Radiator undulator period
	$B_{0\text{Rad}}$	1.11 T	Radiator peak magnetic flux density
	$L_{u\text{Rad}}$	2.4 m	Radiator length
	P_R	1 kW @ $I_P = 2.2$ A	Radiation peak/average power
Soft X-ray (2.7 nm)	σ_z (Rad)	0.5 nm	Linear bunch length at the radiator
	σ_{zy} (Mod)	2.81 μm	Bunch lengthening from ϵ_y at the modulator
	h	1423 m^{-1}	Energy chirp strength
	$\lambda_{u\text{Mod}}$	0.6 m	Modulator undulator period
	$B_{0\text{Mod}}$	0.028 T	Modulator peak magnetic flux density
	$L_{u\text{Mod}}$	1.8 m	Modulator length

(continued)

Table 6.2 (continued)

Parameter	Value	Description
$P_L(Z_R = \frac{L_u}{3})$	1 MW	Modulation laser power
$\lambda_R = \frac{\lambda_u}{100}$	2.7 nm	Radiation wavelength
b_{100}	0.049	Bunching factor
$\lambda_u\text{Rad}$	1 cm	Radiator undulator period
$B_{0\text{Rad}}$	0.86 T	Radiator peak magnetic flux density
$L_u\text{Rad}$	2 m	Radiator length
P_R	1 kW @ $I_P = 5$ A	Radiation peak/average power

circumference of 100 m should be feasible. This compact high-power EUV radiation source is promising to fulfill the urgent need of EUV lithography for high volume manufacture, and also serve the future lithography like Blue-X which invokes 6.x nm-wavelength light source. Such an SSMB-based high-power soft X-ray photon source could be of great value for fundamental science like high-resolution angle-resolved photoemission spectroscopy and can also bridge the water window gap.

References

1. Chao A (2022) Focused laser, unpublished note
2. Li Z, Deng X, Pan Z, Tang C, Chao A (2023) A generalized longitudinal strong focusing storage ring, Generalized longitudinal strong focusing in a steady-state microbunching storage ring, submitted

Printed in the United States
by Baker & Taylor Publisher Services